Go Figure!

Using Math to Answer Everyday Imponderables

Clint Brookhart, P.E.

CB
CONTEMPORARY BOOKS

Library of Congress Cataloging-in-Publication Data

Brookhart, Clint.
    Go figure! :   using math to answer everyday imponderables / Clint
Brookhart.
        p.   cm.
    ISBN 0-8092-2882-3
    1. Mathematics—Popular works.   I. Title.
QA93.B75   1998
519—dc21   97-48461
                                                          CIP

The mathematical principles and the approaches to problem solutions presented in this book
are valid; however, the author assumes no responsibility for the accuracy of the scientific and
statistical data that appear in various sections.

Interior design by Precision Graphics

Published by Contemporary Books
A division of NTC/Contemporary Publishing Group, Inc.
4255 West Touhy Avenue, Lincolnwood (Chicago), Illinois 60646-1975 U.S.A.
Copyright © 1998 by NTC/Contemporary Publishing Group, Inc.
Printed in the United States of America
International Standard Book Number: 0-8092-2882-3

18   17   16   15   14   13   12   11   10   9   8   7   6   5   4   3   2   1

Thanks to Kitty

Contents

PART II: Workings of Time and Space   **75**

PART III: Did You Know?   **115**

Preface

"It is therefore to no purpose to discuss the uses of
knowledge—man wants to know, and when he ceases to do
so he is no longer man."

Fridtjof Nansen, polar explorer

As Bruce Merserve and Max Sobel observed in their *Introduction
to Mathematics* (Prentice-Hall), "Many people study mathematics
just for fun! These individuals would rather solve a mathematical
puzzle than read a book, watch television, or go to a movie.
Admittedly, not everyone has, or can have, this type of disposi-
tion. On the other hand, most of us use mathematical concepts
in a variety of ways but have never been given an opportunity to
explore some of the more interesting aspects of mathematics."

This author couldn't have said it better!

Numbers are around us everywhere. Newcomers to
mathematics will find that dealing with them will be very impor-
tant in molding many careers. For fields of business, social
sciences, health fields, and particularly engineering and science,
mathematics is an invaluable support ingredient.

Unfortunately, many students report that dull, difficult problems are being "stuffed down their throats." So many times exercises totally unrelated to their interests turn students away rather
than attract them to mathematics. It doesn't have to be that way!
There is excitement, beauty, simplicity, and particularly logic in
the field of mathematics.

Theoni Pappas, a well-known author of books popularizing mathematics, has said that her writings are committed to demystifying mathematics and helping eliminate the elitism and fear associated with it. This author would like to be included with those, such as Ms. Pappas, who believe that mathematics is not only a most interesting discipline but also an indispensable part of everyday life.

I anticipate that people with a variety of backgrounds will use this book. If you remember your background in high school mathematics (certain fundamentals such as scientific notation, exponentials, logarithms, and geometric relationships), you will have little difficulty in understanding the material; if you have had little math, or have been away from math for a while, you will want to spend some time brushing up on fundamentals by reviewing the Appendixes.

If *Go Figure!* serves to inspire just one individual to develop an insatiable desire to explore the vast opportunities presented in the field of mathematics, then its purpose will have been fulfilled. May your reading of this book excite the imagination and provide many hours of enjoyment!

Go Figure!

Scientific Calculator Applications

A number of the problem examples in this book include solution instructions using a handheld scientific calculator. They are set off as boxed text and illustrate how simple it is to obtain an answer to an otherwise tedious exercise.

Two kinds of logic are commonly used in pocket calculators: reverse Polish logic and direct algebraic logic. The former avoids the use of parentheses and is highly efficient once you learn the rules. Direct algebraic logic uses parentheses and mimics the procedures of ordinary algebra. For this reason, the scientific calculator operations in this book make use of a typical algebraic calculator.

The key symbols and the keystroke sequences shown on the facing page are common to most calculators made by many manufacturers, but it is assumed that your calculator has at a minimum one memory, two levels of parentheses, log and power functions, and scientific notation. Keystroke sequences shown may be slightly different from those of your calculator; however, your owner's manual should clarify these differences.

The Display

`ON/C`	On and Clears Display
`OFF`	Off and Power Saver
`0` – `9` `.` `+/-`	Data Entry
`+` `−` `X` `÷` and `=`	The Basics

AOS: The Algebraic Operating System

`(` `)`	Parentheses
`EE↓`	Scientific Notation and the Exponential Shift
`nCr` `nPr`	Combinations and Permutations
`1/x`	Inverse Function or Reciprocal
`STO` `RCL` `SUM` `EXC`	Memory Storage Keys
`x²` `√x`	Square and Square Root
`yˣ` and `INV` `yˣ`	Powers and Roots
`K`	Calculations with a Constant
`π`	Pi
`n!` or `x!`	Factorials
`%`	Percent
`DRG`	Angular Measure
`sin` `cos` `tan`	Trigonometric Functions
`log` `lnx`	Logarithms
`INV`	Inverse Key Summary

Adapted from "Understanding Calculator Math" (Developed for Radio Shack by Texas Instruments Learning Center)

Note: More recent models substitute `2nd` for `INV`; both perform identical operations.

Go Figure!

Go Figure!

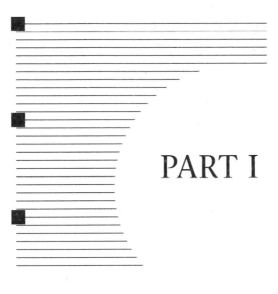

PART I

Something for Everyone

How Far Can You See?

How Far Is the Horizon from a Height?

Many are curious about how far they can see from a great height. For instance, from 1,440 feet (the top floor of the John Hancock Building in Chicago), how far away is the horizon?[1]

There is a formula to approximate the distance to the horizon that is accurate to 0.5 percent at 180,000 feet above the ground.

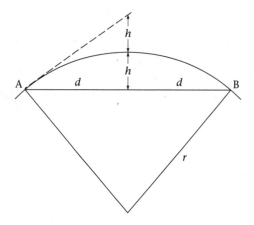

In the figure, h is the height of the building. The straight line from A to B is a chord of a circle, and r is the radius of the earth. The dashed line is the observer's line of sight, and d is the distance the observer can see to the horizon. Because the length of the chord from A to B is $2d$, it can be shown that:

---

[1]From the top of the John Hancock building, the distance is about $1.225\sqrt{1440} \approx 46.5$ miles.

$$(2d)^2 \approx 8rh$$

Thus, the distance is $d \approx \sqrt{2rh}$. The radius of the earth, r, is 3,963 miles. The height, h, of the building is given in feet, so use the conversion factor (1 mile = 5,280 feet).

$$d \approx \sqrt{2(3{,}963 \text{ mi}) \times \frac{1 \text{ mi}}{5{,}280 \text{ ft}} \times h \text{ ft}}$$
$$\approx \sqrt{1.501 \text{ mi}^2 \times h}$$
$$\approx 1.225\sqrt{h} \text{ miles, provided } h \text{ is in feet}$$

Example: How far is the horizon from an airplane flying at 40,000 feet?

$$d \approx 1.225 \sqrt{h} \approx 1.225 \times 200 \approx 245 \text{ miles}$$

Example: From the 12th floor of your ocean-front hotel, which is about 150 feet above sea level, how far away is the horizon?

$$d \approx 1.225\sqrt{150} \approx 15 \text{ miles}$$

Press   1 5 0 √x̄ X 1 . 2 2 5 =
Display   15.003125

Baby, It's Cold Outside!
Effects of Temperature and Wind

Heavy clothing, a building furnace, and a car heater can help keep you warm in cold weather, but if you're outside too long or you're not adequately dressed, you may develop chilblains or frostbite on the nose, ears, cheeks, and toes. The effect of cold on your body is more severe if you are also exposed to wind. The *windchill equivalent temperature* is calculated by using a formula that relates temperature and wind speed. For example, if you're exposed to air that is 5°F and wind that is blowing at 15 miles per hour, the effect on your body is the same as if you were exposed to –25°F air with no wind.

The windchill equivalent temperature, or *windchill index*, can be calculated using the following formula:

$$T_{wc} = 91.4 - \frac{(10.45 + 6.686\sqrt{v} - 0.447v) \times (91.4 - T_f)}{22}$$

T_{wc} is the windchill index, v is the wind velocity in miles per hour, and T_f is the temperature of the air in degrees Fahrenheit. For example, if the wind velocity is 25 miles per hour and the air temperature is 5°F, the windchill index is –37°F.

As wind speeds increase from 40 miles per hour, the chilling effect lessens. For example, a wind speed of 50 miles per hour combined with a temperature of 5°F has the same chilling effect on your body as a temperature of –48°F when the wind is calm. If the temperature remains at 5°F and the wind speed increases to 70 miles per hour, the effect is the same as –47°F.

At wind speeds of 4 miles per hour or less, the windchill temperature is the same as the actual air temperature.

Go Figure!

Windchill Index

Wind Velocity, mi/hr	\ Air Temperature, °F → 35	30	25	20	15	10	5	0	−5	−10	−15	−20	−25	−30	−35	−40	−45
4	35	30	25	20	15	10	5	0	−5	−10	−15	−20	−25	−30	−35	−40	−45
5	32	27	22	16	11	6	0	−5	−10	−15	−21	−25	−31	−36	−42	−47	−52
10	22	16	10	3	−3	−9	−15	−22	−27	−34	−40	−46	−52	−58	−64	−71	−77
15	16	9	2	−5	−11	−18	−25	−31	−38	−45	−51	−58	−65	−72	−78	−85	−92
20	12	4	−3	−10	−17	−24	−31	−39	−46	−53	−60	−67	−74	−81	−88	−95	−103
25	8	1	−7	−15	−22	−29	−37	−44	−51	−59	−66	−74	−81	−88	−96	−103	−110
30	6	−2	−10	−18	−25	−33	−41	−49	−56	−64	−71	−79	−86	−93	−101	−109	−116
35	4	−4	−12	−20	−27	−35	−43	−52	−58	−67	−74	−82	−89	−97	−105	−113	−120
40	3	−5	−13	−21	−29	−37	−45	−53	−60	−69	−76	−84	−92	−100	−107	−115	−123
45	2	−6	−14	−22	−30	−38	−46	−54	−62	−70	−78	−85	−93	−102	−109	−117	−125

Region bands (from warmer to colder): Cold — Very Cold — Bitter Cold — Extreme Cold

Air Temperature, °F

Wind Velocity, mi/hr

Source: National Atmospheric and Oceanic Administration

Hey! Cool It!
Temperature-Humidity Index

When the relative humidity and air temperature are both high, perspiration on our skin evaporates more slowly and we sense a higher temperature than the temperature of the air. The familiar saying "It's not the heat; it's the humidity" is not strictly true because the oppressive feeling we associate with hot, humid weather is caused equally by high temperature and high humidity.

An objective index of the temperature we feel is based on air temperature and humidity. The temperature we feel is approximated by the *temperature-humidity index* (THI), also called the discomfort index.

In the following formula,[1] T_f is the air temperature in degrees Fahrenheit and H is the relative humidity in percent.

$$THI = 36 + \frac{(23 + 0.22H) \times (T_f - 48)}{22}$$

Example: If the humidity (H) is 80% and the temperature (T_f) is 90°F, the discomfort index is:

$$THI = 36 + \frac{(23 + 0.22 \times 80) \times (90 - 48)}{22} = 113°F$$

Press $\boxed{0}\,\boxed{\cdot}\,\boxed{2}\,\boxed{2}\,\boxed{\times}\,\boxed{8}\,\boxed{0}\,\boxed{+}\,\boxed{2}\,\boxed{3}\,\boxed{=}\,\boxed{\times}\,\boxed{(}\,\boxed{9}\,\boxed{0}\,\boxed{-}$
               $\boxed{4}\,\boxed{8}\,\boxed{)}\,\boxed{\div}\,\boxed{2}\,\boxed{2}\,\boxed{+}\,\boxed{3}\,\boxed{6}\,\boxed{=}$

Display   113.50909

---

[1] The author developed this formula from several that are in use.

The table gives the THI for a wide range of relative humidity and air temperature values.

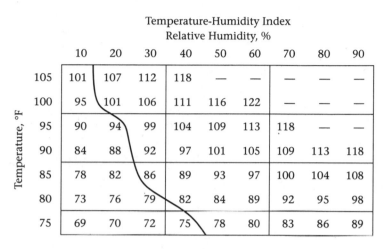

Temperature-Humidity Index
Relative Humidity, %

Temperature, °F	10	20	30	40	50	60	70	80	90
105	101	107	112	118	—	—	—	—	—
100	95	101	106	111	116	122	—	—	—
95	90	94	99	104	109	113	118	—	—
90	84	88	92	97	101	105	109	113	118
85	78	82	86	89	93	97	100	104	108
80	73	76	79	82	84	89	92	95	98
75	69	70	72	75	78	80	83	86	89

To the right of the curved line, the index is higher than the temperature.

A skier tells you that at 20°F he is much more comfortable skiing in Colorado than in northern New York because the air is drier in Colorado. The windchill index does not take humidity into account and, as we know, the lower the humidity, the more comfortable we are in cold weather.

In the summer, a jogger welcomes a slight breeze, which provides some relief from the high temperature and high humidity. The temperature-humidity index does not reflect the cooling effect of the breeze.

A formula that relates the effects of wind velocity, humidity, and temperature needs to be researched and developed.

Easy Volumes
One Formula Simplifies It All

You don't need to memorize numerous formulas to figure out the volume of different solid shapes. You can use one formula, called the *prismoidal formula,* to find the volume of any solid with two parallel faces if the faces are polygons having the same number of sides. Such a solid is called a *prismoid.*

The volume of the prismoid shown here is calculated using the formula given below it.

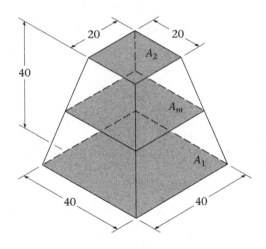

$$V = \left(\frac{A_1 + 4A_m + A_2}{6}\right)h$$

A_1 is the area of the base, A_m is the area of a section halfway up the prismoid, A_2 is the area of the top of the prismoid, and h is the vertical height of the prismoid.

Go Figure!

If you cut off the top of a prismoid parallel to its base, you have the *frustum of a pyramid,* which is one kind of prismoid. You can use the prismoidal formula to find the volume. To find the area of the middle section, you first need to find the length of the sides of the middle section, which is the average of the sides of the base and those of the top.

Base $\qquad A_1 = 40 \times 40 = 1,600$

Middle $\qquad A_m = \left(\dfrac{40 + 20}{2}\right)^2 = 30^2 = 900$

Top $\qquad A_2 = 20 \times 20 = 400$

Height $\qquad h = 40$

$$V = \left(\frac{1,600 + 4(900) + 400}{6}\right) \times 40 = 37,333.3$$

This value agrees with the one obtained from the formula for the volume of the frustum of a pyramid:

$$V = (A_1 + A_2 + \sqrt{A_1 A_2})(h/3)$$
$$= (1,600 + 400 + \sqrt{1,600 \times 400})\frac{40}{3} = 37,333.33$$

The prismoidal formula can even be applied to volumes of curved shapes. For a sphere, such as the one shown here, the top and base areas are both zero; the area of the middle section is the area of the circle that is the equator.

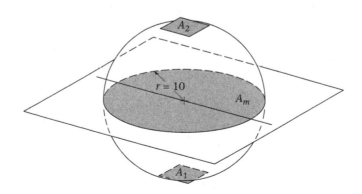

$$A_1 = 0$$
$$A_m = \pi\, r^2 = 100\pi$$
$$A_2 = 0$$
$$h = 2r20$$
$$V = \left(\frac{0 + 4(100\pi) + 0}{6}\right)(20) = 4{,}188.79$$

This number agrees with the one obtained from the formula for the volume of a sphere:

$$V = \frac{4}{3}\pi\, r^3 = \frac{4}{3} \times 1{,}000\pi = 4{,}188.79$$

To find the volume of an ellipsoid, find the area of the middle section by using the formula for the area of an ellipse, $A = \pi ab$, where a is the semi-major axis and b is the semi-minor axis.

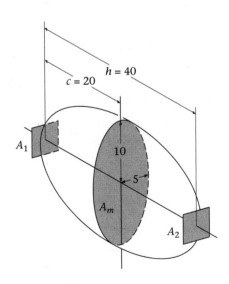

Go Figure!

$$A_1 = 0$$

$$A_m = \pi ab = 10 \times 5 \times \pi = 157.08$$

$$A_2 = 0$$

$$h = 40$$

$$V = \left(\frac{0 + 4(157.08) + 0}{6}\right)(40) = 4{,}188.8$$

Using the formula for the volume of an ellipsoid, we get

$$\frac{4}{3}\pi abc = \frac{4}{3}\pi(10)(5)(20) = 4{,}188.8$$

which agrees with the number obtained with the prismoidal formula.

The prismoidal formula is particularly useful for approximating the volume of irregular solids, such as the volume of earthworks when building highways.

Looking Back
How Many Direct Ancestors Do You Have?

You can find the number of your parents, grandparents, great-grandparents, and so on, going back any number of generations by computing the sum of a geometric series. For example, the expression

$$\sum_{i=1}^{n} 2^i$$

means the sum of powers of 2 as the exponent goes from 1 to n. There are n terms in this sum and the last exponent is n. Written out, this expression is $2 + 4 + 8 + 16 + \cdots + 2^n$.

To find the number of ancestors for four generations, $n = 4$, so:

$$\sum_{i=1}^{n} = 2^i = 2^1 + 2^2 + 2^3 + 2^4$$
$$= 2 + 4 + 8 + 16 = 30$$

Computing the sum of the terms in this geometric series for large numbers (n) can be laborious. Fortunately, the following formula is much easier:

$$\sum_{i=1}^{n} 2^i = a_1 \frac{r^n - 1}{r - 1}$$

In this formula, a_1 is the first term in the series (2), and r is the common ratio of the terms (also 2). Each term is twice the term before.

Now let's use the formula to check our answer. Do you really have thirty direct ancestors in the previous four generations?

$$\sum_{i=1}^{4} 2^i = 2\frac{2^4 - 1}{2 - 1} = 2\frac{16 - 1}{1} = 2(15) = 30$$

If you use a calculator, here are the steps.

Press $\boxed{2}\ \boxed{y^x}\ \boxed{4}\ \boxed{-}\ \boxed{1}\ \boxed{=}\ \boxed{\times}\ \boxed{2}\ \boxed{=}$

Display   **30**

If each generation spans about twenty-five years, twelve generations span 12×25 or about 300 years. How many of your ancestors lived during this time?

$$\sum_{i=1}^{12} 2^i = 2\frac{2^{12} - 1}{2 - 1} = 2\frac{4,096 - 1}{1} = 8,190$$

As many as 8,190 people played a role in your being here today, fewer if any of those 8,190 people had common ancestors themselves.

You can use the formula for the sum of a geometric series to solve the following example.

Example: A worker was offered a month's employment at 2 cents for the first day, 4 cents for the second day, then 8 cents, 16 cents, 32 cents, and so forth. What are the total wages for four 5-day weeks (the twenty working days in a month) using the formula, $a_1 = 2$, $r = 2$, and $n = 20$?

$$\sum_{i-1}^{20} 2^i = 2\frac{2^{20} - 1}{2 - 1} = 2(1,048,576 - 1)$$

$$= 2,097,150 \text{ cents or } \$20,971.50$$

Not bad for twenty days' work!

Looking for a Date?
Carbon Dating and the Shroud of Turin

Carbon dating is a method for determining the age of prehistoric organic material by measuring its radioactivity. The technique was developed by an American chemist, Dr. Willard F. Libby (1908–1980), in the late 1940s. All living things absorb carbon 14 from carbon dioxide in the atmosphere. After an animal or plant dies, the amount of carbon 14 present begins to break down by releasing particles at a uniform rate. Every 5,730 years, half of the carbon 14 has decayed. If the remaining carbon 14 is measured and compared to that of a living sample, the elapsed time since the animal or plant died can be determined. For example, if a prehistoric axe handle has one-fourth as much carbon 14 as a living tree, the tree the handle was made from must have lived $2 \times 5,730$ or 11,460 years ago.

The Shroud of Turin, an ancient piece of linen bearing a human image, was widely believed to be the burial cloth of Jesus Christ. The shroud, which is 14 feet 3 inches long and 3 feet 7 inches wide, is mentioned in documents as early as A.D. 1354, but its earlier history is obscure. Since 1578, it has been preserved and venerated in the cathedral of Turin, Italy.

When first photographed in 1898, the image on the shroud, that of the front and back of a crucified man about 6 feet tall, appeared as a negative rather than a positive. Scientific tests conducted for the Vatican in 1988 concluded that the shroud itself dates back no earlier than the year 1260.

The formula $T = -8,267 \ln x$ is sometimes used to predict the age T (in years) of organic material, where x is the percentage (expressed as a decimal) of carbon 14 still present in the fossil. (See Appendix D for natural logarithms.)

Chemical analysis performed in 1988 found that 91.6 percent of the carbon 14 remained in linen fibers from the Shroud of Turin.

$$T = -8{,}267 \ln 0.916$$
$$T = -8{,}267 \,(-0.08773891)$$
$$T = \text{about } 725 \text{ years}$$

You can use your calculator to find the age of the Shroud of Turin.

Press   [0] [.] [9] [1] [6] [Inx] [+/-] [X] [8] [2] [6] [7] [=]

Display   725.3

As the product of two negative numbers is a positive number, the [+/-] is needed only once.

Since the tests were performed in 1988 and the age of the shroud is about 725 years, it dates from about the year 1988 − 725 = 1263, according to this formula.

Carbon dating can be used for materials from a few hundred years old to about fifty thousand years old. If chemical analysis could detect as little as 0.3 percent of carbon remaining in an organic sample, what would the age T of the sample be?

$$T = -8{,}267 \ln 0.003$$
$$T = -8{,}267 \,(-5.809)$$
$$T = 48{,}024 \text{ years}$$

California Earthquakes

Three Mini Big Ones in Four Years!

Californians grow up with earthquakes. Most quakes are small, less than magnitude 5. The larger ones are usually off the coast or in the desert.

In the seventy-five years preceding the great San Francisco earthquake of 1906, sixteen earthquakes with magnitudes greater than 6 struck the Bay Area. For nearly as many years after that, only one quake as large struck, a 6.5 in 1911. Now seismic activity is increasing again in the region: four earthquakes with magnitudes greater than 5.7 have hit since 1979. Likewise, southern California has been averaging one quake of that size every year since 1986.

The earth's surface is riddled with cracks called faults. An earthquake occurs when forces deep in the earth rupture a crack. The uneasy San Andreas Fault, which bisects California from the Mexican border to the northern coast, is becoming more active. The San Andreas Fault drives most of California's seismic turmoil. The main fault connects fault segments running through the state, including a zone of parallel, branching faults as long as 100 miles.

The U.S. Geological Survey (USGS) conservatively gives the Bay Area a 67 percent chance of having at least one 7.0 or stronger quake from those faults in the next thirty years. David Swartz of the USGS suggests that the probability is closer to 90 percent. Each one could be a 30-billion-dollar quake, estimates Bill Bakun of the USGS.

Earthquakes are measured on the Richter scale. Charles Richter, a California seismologist, invented it in 1935 because he was tired of journalists asking him to compare the size of earthquakes. The Richter scale uses a seismograph to measure seismic waves, the form of energy released when rocks deep within the earth are disrupted.

Go Figure!

Many newspaper stories have been published recently about the "Big One." The Big One doesn't have to be a 7.5 or 8.0. Ron Egachi, a hazard-assessment specialist at the engineering firm EQE International, says, "Just one magnitude 7 quake under Los Angeles could kill 5,000 people in Los Angeles County if it struck during the day. Another 50,000 could be injured and 500,000 might be left homeless. Damage to buildings could run 50 billion dollars, with another 25 billion in repairs to such regional infrastructures as roads, bridges, and utility lines."[1]

Since early 1989, California has had three "mini Big Ones." These quakes were not so "mini" in terms of destruction and loss of life.

On October 17, 1989, the Loma Prieta earthquake, centered about 70 miles south of San Francisco, measured 7.1 on the Richter scale. The brunt of Loma Prieta was felt in Santa Cruz, which lost 40 percent of its downtown buildings. One section of the upper deck of the San Francisco–Oakland Bay Bridge collapsed, closing the bridge for a month. (This bridge and the Golden Gate Bridge are now undergoing a seismic retrofit at a combined cost of 450 million dollars.) Four hundred buildings crumbled to the ground. Seven hundred more were badly damaged. In that single day, the Loma Prieta repair bill amounted to 5 billion dollars, and 62 lives were lost.

On June 28, 1992, the Mojave Desert town of Landers, 100 miles east of Los Angeles, suffered a 7.3 earthquake, which rumbled as far away as Montana. Fortunately, the Landers quake struck a relatively unpopulated area. The quake began several miles underground in the Johnson Valley Fault. Roads near Landers shifted horizontally as much as 10 feet, and dozens of aftershocks were recorded during the next 10 days.

The third of the mini Big Ones, the most devastating California earthquake since 1906, struck at 4:31 A.M., January 17, 1994. (Is there something about the seventeenth day of the month? The Loma Prieta happened on October 17.) The earth's

---

[1] Adapted from *National Geographic Magazine*, April 1995.

crust snapped miles beneath the community of Northridge, just 14 miles north of Hollywood. The magnitude 6.7 quake killed 60 people. It destroyed or left uninhabitable more than 3,000 homes. It toppled ten highway bridges and closed three major freeways. Its spasms demolished part of a huge shopping mall in Northridge and leveled seven concrete parking structures. The damage was estimated at 20 million dollars!

The Richter scale is a logarithmic scale. The magnitudes of quakes vary logarithmically as the powers of 10. For instance, a 7.3 quake is $10^{7.3}/10^{6.3}$ times greater than a 6.3 quake, about $10^{7.3}/10^{6.3} = 10^{7.3-6.3} = 10$ times as great! Those "little" ones measuring 4.3 are only one-thousandth as large as the 7.3 one: $10^{4.3}/10^{7.3} = 10^{4.3-7.3} = 10^{-3}$ or 1/1,000.

How far away from the origin does the earth's surface vibrate? An approximate formula for West Coast earthquakes is

$$d = \left(\frac{10^{(R+7.5)/2.3} - 34,000}{\pi} \right)^{1/2}$$

where R is the magnitude on the Richter scale and d is the distance in miles. For the 1992 Landers quake, which measured 7.3, the shocks were felt over the following distance:

$$d = \left(\frac{10^{(7.3+7.5)/2.3} - 34,000}{\pi} \right)^{1/2}$$

$$d = \left(\frac{2,721,340 - 34,000}{\pi} \right)^{1/2}$$

$$d = (855,400)^{1/2} = \text{about 925 miles}$$

Use this sequence to evaluate d with the calculator. Note that $x^{1/2}$ is equal to $\sqrt{x}$

Press ⎣7⎦⎣.⎦⎣3⎦⎣+⎦⎣7⎦⎣.⎦⎣5⎦⎣=⎦⎣÷⎦⎣2⎦⎣.⎦⎣3⎦⎣=⎦⎣INV⎦⎣log⎦⎣−⎦
⎣3⎦⎣4⎦⎣0⎦⎣0⎦⎣0⎦⎣=⎦⎣÷⎦⎣π⎦⎣=⎦⎣√x⎦

Display   924.9

Go Figure!

Dollars and Sense
High Finance Made Simple

Two areas in finance are of interest to us ordinary folks: loan payments and annuities.

Repaying a Loan

How much are your monthly payments on a loan? Of course, we can use interest tables, but let's do our own figuring! The following formula is simple to apply. If P is the principal or the amount borrowed, n_m is the number of *monthly* payments, and i_m is the *monthly* interest rate, then the amount of each monthly payment is:

$$A = \left(\frac{i_m}{1 - (1 + i_m)^{-n_m}}\right)P$$

For example, you purchase a $20,000 automobile at 12% annual interest to be paid for over 48 months. How much do you pay each month? Here, $P = \$20,000$, $n_m = 48$ payments, and the monthly interest rate is $i_m = 0.12/12 = .01$.

$$A = \left(\frac{0.01}{1 - (1.01)^{-48}}\right)\$20,000$$

$$= \left(\frac{0.01}{1 - \dfrac{1}{1.612226}}\right)\$20,000$$

$$= \left(\frac{0.01}{0.37974}\right)\$20,000$$

$$= \$526.68$$

Remember that $a^{-x} = 1/a^x$. Use your calculator to evaluate 1.01^{-48} first.

Press $\boxed{1}\boxed{\cdot}\boxed{0}\boxed{1}\boxed{y^x}\boxed{4}\boxed{8}\boxed{=}\boxed{1/x}\boxed{-}\boxed{1}\boxed{=}\boxed{+/-}\boxed{1/x}$

       $\boxed{\times}\boxed{0}\boxed{\cdot}\boxed{0}\boxed{1}\boxed{\times}\boxed{2}\boxed{0}\boxed{0}\boxed{0}\boxed{0}\boxed{=}$

Display   526.68

If you paid this amount for 48 months, what total interest would you have paid? Find the total of the payments and subtract the principal.

$$I = (48)(526.68) - 20{,}000$$
$$= 25{,}280.64 - 20{,}000$$
$$= \$5{,}280.64$$

The amount of interest is approximately 26.5% of the principal!

After you make some of the payments, you may want to know the balance due on the principal. Suppose in the previous example you made 23 payments on your $20,000 loan and wonder if you can pay off the loan now. What is the payoff amount? If A is the monthly payment, n_t is the total number of payments (48), N is the number of payments you have made (23), and i_m is the interest rate per month, the balance due is:

$$B = A\left(\frac{1 - (1 + i_m)^{N-n_t}}{i_m}\right)$$

For your auto loan, $A = \$526.68$. Now find B.

$$B = 526.68\left(\frac{1 - (1.01)^{23-48}}{.01}\right)$$
$$= 526.68\left(\frac{1 - (1.01)^{-25}}{0.01}\right)$$

Go Figure!

$$= 526.68 \left(\frac{1 - 0.77977}{.01} \right)$$
$$= 526.68\,(22.023156)$$
$$= \$11{,}599.16$$

Would you like to know how much interest you have paid in these 23 payments? Use the formula $I = NA - (P - B)$ to find the amount of interest.

$$I = (23)(526.68) - (20{,}000 - 11{,}599.16)$$
$$= 12{,}113.64 - 8{,}400.84$$
$$= \$3{,}712.80$$

Investing in Annuities

The same formula can be applied to the reverse situation, called an *annuity*, a single amount invested at a fixed interest rate to provide for periodic payments over the term of the annuity.

For instance, $100,000 is invested today at 6% annual interest. What annual payment will be provided if the term of the annuity is 20 years?

$$A = \left[\frac{0.06}{1 - (1 + 0.06)^{-20}} \right] 100{,}000$$
$$= (0.06)(1.453076)(100{,}000)$$
$$= 8718.46$$

Say your retirement years are imminent. You want to retire now and have a certain annual income of x dollars per year for the next y years. By rearranging the formula for monthly payments, we get the following relation, in which P is the annual income, A is the amount you must invest now, n_a is the number of years of retirement, and i_a is the annual interest rate.

$$P = \left[\frac{1 - (1 + i_a)^{-n_a}}{i_a} \right] A$$

The following table gives values of A for five given values of P, the annual income using 5% as the after-tax interest rate. For instance, suppose you want to retire now on $30,000 per year for the next 30 years. The table shows that you must invest $461,174 today.

Annuity Amounts for Desired Annual Income

Years (y)	Annual Income P				
	$20,000	$25,000	$30,000	$35,000	$40,000
15	207,593	259,491	311,390	363,288	415,186
20	249,244	311,555	373,866	436,177	498,488
25	281,879	352,349	422,818	493,288	563,758
30	307,449	384,311	461,174	538,036	614,898
35	327,484	409,355	491,226	573,097	654,968

Calculating Growth

Another related area of finance is also useful to us small investors. For example, if your investment has appreciated three- or fourfold over n years, you would like to know the annual rate of growth in percent. The following formula gives the interest rate i in terms of n, the number of years, and N, as in N-fold:

$$i = [10^{(\log N)/n} - 1]100$$

Suppose an investment has appreciated sixfold over 8 years. What is the annual percentage gain? Use the formula with $N = 6$ and $n = 8$.

$$\begin{aligned} i &= [10^{(\log 6)/8} - 1]100 \\ &= (10^{0.77815/8} - 1)100 \\ &= (10^{0.09727} - 1)100 \\ &= (1.251 - 1)100 = 25.1\% \end{aligned}$$

The value $10^{(\log 6)/8}$ is found easily with your calculator, if you realize that 10^x is the inverse of the log function. We first calculate $\frac{\log 6}{8}$, then use the inverse key and log key to determine $10^{(\log 6)/8}$.

Press $\boxed{6}\,\boxed{\log}\,\boxed{\div}\,\boxed{8}\,\boxed{=}\,\boxed{\text{INV}}\,\boxed{\log}\,\boxed{-}\,\boxed{1}\,\boxed{=}\,\boxed{\times}\,\boxed{1}\,\boxed{0}\,\boxed{0}\,\boxed{=}$

Display   25.1

Example: The price of a gallon of gasoline in 1994 was about $1.29. In 1928 it was $0.17. What is the annual inflation rate for the price of gasoline? Use the formula with $N = 1.29/0.17 = 7.59$, and $n = 1994 - 1928 = 66$ years.

$$
\begin{aligned}
i &= [10^{(\log 7.59)/66} - 1]100 \\
&= (10^{0.88024/66} - 1)100 \\
&= (1.0312 - 1)100 = 3.12\%
\end{aligned}
$$

The price of gasoline has not risen as much as we might have guessed. The average inflation rate of all goods and services was about 4% over those 66 years. At that rate, the 1994 price of gasoline would be $1.04^{66} \times 0.17$ or $2.26 per gallon.

The Making of a Star
Doing It the Old-Fashioned Way

In the old days, before protractors were available to measure angles, only a compass and a straightedge were used to make geometric constructions. Here we shall draft a five-point star with only these instruments.

Step 1. Bisect the radius $\overline{OB}$ (see Figure 1). Place the compass point at O; draw arcs that lie above and below the circle. Now place the compass point at B and draw arcs that intersect the first two, at R and S. Connect R and S. This line is the bisector of $\overline{OB}$.

Step 2. Place the compass point at A (the midpoint of $\overline{OB}$) and set the radius for the length AC. Draw arcs at C and D.

Step 3. Set the radius of the compass for the length CD. Place the compass point at C and mark off point E on the circle. Proceeding around the circle (see Figure 2), mark off points F, G, and H.

Step 4. Draw the five-point star by connecting points as shown in Figure 2.

Now Prove It!

Do the five points of your star truly and accurately divide the 360 degrees of the circle into five equal arcs?

Using Figure 3, let the radius of the circle ($\overline{OB}$) equal one unit so that OAC is a right triangle. Recall that A is the midpoint of $\overline{OB}$. Use the Pythagorean theorem:

Go Figure!

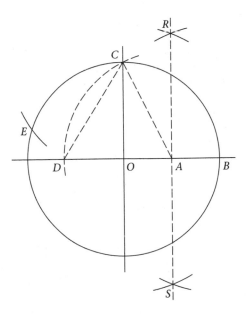

Figure 1

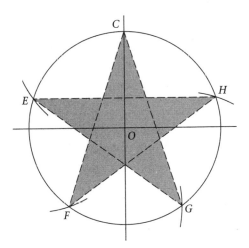

Figure 2

$$AC^2 = 0.5^2 + 1.0^2$$
$$AC = \sqrt{0.5^2 + 1.0^2} = \sqrt{1.25}$$

Because $AC = AD$:

$$OD = \sqrt{1.25} - 0.5$$

Then, by the Pythagorean theorem:

$$CD^2 = 1.0^2 + (\sqrt{1.25} - 0.5)^2 = 1.381966$$
$$CD = \sqrt{1.381966} = 1.17557$$

CD is one side of a regular pentagon inscribed in a circle of radius 1, whose center is A.

To show that the points C, E, F, G, and H do divide the circle into five equal arcs, we must prove that CE in Figure 4 is one side of an inscribed pentagon. The central angle of a pentagon is $2\pi/5$, so:

$$\angle COZ = \frac{1}{2} \times \frac{2\pi}{5} = \frac{\pi}{5}$$

Using trigonometry and a Taylor series, the ratio of CZ to OC can be evaluated.

$$\frac{CZ}{OC} = 0.587785$$

Now $OC = 1$ and:

$$CE = 2 \times CZ$$
$$2 \times 0.587785 = 1.17557$$

Because 1.17557 is the length of the side of a pentagon inscribed in a circle of radius 1, CE is the side of a regular pentagon ($CD = CE$ by construction). Therefore, the five points do divide the circle into five equal arcs. QED!

Go Figure!

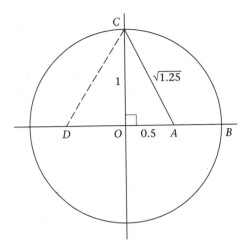

Figure 3

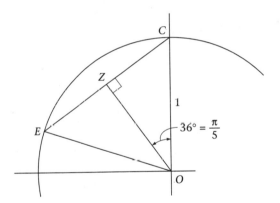

Figure 4

Ups and Downs

Exponential Growth and Decay

Many areas of science, sociology, and business may be better understood and explored using the principles of exponential growth and decay. Such areas include analyzing or predicting the rate of growth (or decline) in local or world populations, bacteria, or a company's retail sales.

We can use the following formula to determine exponential change:

$$N(t) = N_0 \times e^{kt}$$

$N(t)$ is a quantity whose rate of change is proportional to the quantity of time t, N_0 is the initial value, and k is the constant of proportionality.

Exponential changes can also be illustrated graphically. When something grows exponentially (represented here by a), its growth follows an ever-increasing curve, as shown below. Where $f(t) = N(t)$ for a specific N, $f(t) = a^{kt}$ for $k > 1$.

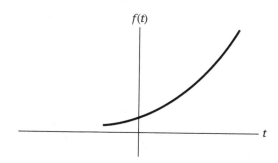

Conversely, an exponential decline follows an ever-decreasing curve, as shown below. Here, $f(t) = a^{-kt}$ for $k < 1$.

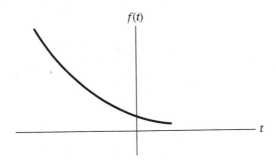

Example: The population of the United States was 248.3 million in 1990. Since that time, it has grown at the rate of 0.76% per year. If this rate continues, what is the predicted U.S. population for the year 2050?

Using our formula, $t = 60$ years, $N_0 = 248.3$ (million), and $k = 0.76\%$ or 0.0076.

$$
\begin{aligned}
N(t) &= 248.3e^{0.0076(60)} \\
&= 248.3e^{0.456} \\
&= 248.3 \times 1.5778 \\
&= 392
\end{aligned}
$$

Thus, the U.S. population in 2050 is predicted to be 392 million.

A scientific calculator makes finding the solution to this exercise fairly simple. If you don't have an e^x key, recognize that e^x is the inverse of the natural logarithm, ln. Just follow this sequence:

Press   0 · 0 0 7 6 X 6 0 = INV ln X 2 4 8 · 3 =

Display   391.76

Example: When we speak of a cohort in fishing terms, we are referring to the total number of fish in one yearly reproduction cycle. Suppose you wanted to estimate the percentage of a cohort of Pacific halibut that would still be alive after six years. Using our formula, $N(t)$ is the number of fish still alive, N_0 is the initial size of the cohort, t is 6 years, and k is assumed to be a rate of decline of 20% per year.

$$N(t) = N_0 e^{-0.20(6)}$$
$$= N_0 e^{-1.2} = 0.301 N_0$$

Thus, the percentage of the initial cohort still alive would be 30%.

Go Figure!

Patterns in Arithmetic

The Ubiquitous Number Nine

Mathematicians love to search for patterns and generalizations, whether in arithmetic, algebra, or geometry. Finding patterns is not only interesting, but may help you better understand mathematics as a whole.

The patterns discussed here all involve the number nine. Even if some of them are familiar to you, a brief review never hurts!

Casting Out Nines: Multiplication

This method, generally taught in grade school, is useful for checking the results of multiplication. Say you wish to check the accuracy of 31,256 × 8,427 = 263,394,312. You first add the digits of the multiplicand, multiplier, and product to get 17, 21, and 33, as shown below. Because each of these numbers is greater than 9, you add the digits of the individual sums to get 8, 3, and 6.

	Sum of Digits	**Sum of Digits**	
31,256	17	8	
× 8,427	21	3	8 × 3 = 24
263,394,312	33	6	2 + 4 = 6

Using the numbers in the last column, you now multiply those of the multiplicand and the multiplier (8 × 3 = 24). The sum of 2 and 4 is 6 (as is the final sum for the product), so the answer is correct.

Casting Out Nines: Addition

The underlying principle to this method is the same as the one we have already discussed for multiplication. Say you have added together a series of numbers, and you want to check the accuracy of the sum. Because each of the numbers is greater than 9, you first add the digits of the individual addends and then add the digits of each of the sums, as shown.

	Sum of Digits	Sum of Digits
4,378	22	4
2,160	9	9
3,872	20	2
+ 1,085	14	5
11,495	20	2

The sum of the digits in the product is 20, which breaks down to $2 + 0 = 2$. If you add together the final sums for the addends ($4 + 9 + 2 + 5$), the answer again is 20 ($2 + 0 = 2$). Therefore, the answer 11,495 is correct.

Tricks of Division

There are many interesting "nine" patterns to be found in the rules of division as well. Did you know that, if a number is diminished by the sum of its digits, the result is always divisible by 9? For example:

$$173,962 - (1 + 7 + 3 + 9 + 6 + 2)$$
$$173,962 - 28 = 173,934$$
$$173,934 \div 9 = 19,326$$

Similarly, a number is divisible by 9 if the sum of its digits is divisible by 9. For example, the sum of the digits in 234,567 is 27, which is divisible by 9. Therefore, it is possible to divide 234,567 by 9 also (234,567 ÷ 9 = 26,063).

A Remarkable Common Property

Choose any number with two or more digits; for example, 76,493. Reverse the digits and subtract the second number from the first (76,493 − 39,467 = 37,026). The difference is always divisible by 9 (37,026 ÷ 9 = 4,114).

An Interesting Pattern

This pattern shows how the number of ones in the group on the right corresponds to the last digit of each equation on the left.

$$123,456 \times 9 + 7 = 1,111,111 \text{ (or 7 ones)}$$
$$1,234,567 \times 9 + 8 = 11,111,111 \text{ (or 8 ones)}$$
$$12,345,678 \times 9 + 9 = 111,111,111 \text{ (or 9 ones)}$$

A related pattern, although it has nothing to do with nines, is also interesting:

$$1 \times 1 = 1 = 1^2$$
$$11 \times 11 = 121 = 11^2$$
$$111 \times 111 = 12,321 = 111^2$$
$$1,111 \times 1,111 = 1,234,321 = 1,111^2$$
$$11,111 \times 11,111 = 123,454,321 = 11,111^2$$

Backward Numbers
How the n *Factorial Works*

We see the n factorial, written as $n!$, throughout mathematical formulas and expressions, particularly in many types of series (the sum of a usually infinite sequence of numbers). Factorial notation can be defined as $n! = n \times (n-1) \times (n-2) \times (n-3) \ldots$ and so on. Because sums in these series increase rapidly, it is useful to be able to approximate when dealing with large values of n.

James Stirling (1692–1770) developed an approximation for $n!$ that is still widely used and unsurpassed in accuracy. It is known as *Stirling's approximation*:

$$n! = \left(\frac{n}{e}\right)^n (2\pi n)^{\frac{1}{2}}$$

Let's see how well Stirling's formula works when $n!$ grows exponentially. For the purpose of example, we will calculate 12!:

$$n! = \left(\frac{12}{2.71828}\right)^{12} (2 \times \pi \times 12)^{\frac{1}{2}}$$
$$= (5.47824 \times 10^7)(8.6832)$$
$$= 4.7569 \times 10^8 \text{ or } 475{,}690{,}000$$

This is a good approximation of the exact number, which is 479,001,600 (it's only 0.696% off).

Next, let's try using a slightly larger number, such as 20:

$$n! = \left(\frac{20}{2.71828}\right)^{20} (2 \times \pi \times 20)^{\frac{1}{2}}$$
$$n! = (2.1612762 \times 10^{17})(11.2099824)$$
$$n! = 2.422787 \times 10^{18}$$

You can also use a handheld calculator that has *n*! capability,

Press   2 0 n! =

Display   2.4329 18

Finally, let's compare the two factorials we computed:

$$\frac{20!}{12!} = \frac{(2.422787 \times 10^{18})}{(4.7569 \times 10^{8})} = 5.1 \times 10^{9}$$

The summation does grow exponentially!

"Horizontaling" a Slope

Slope Correction Measurements Made Easy

When surveyors make horizontal land measurements, they use a steel tape called a chain. During "chaining" (or measuring), they often need to adjust their calculations to compensate for the various dips and inclines in the ground. With these adjustments, the distance of positive and negative slopes can be made "horizontal."

The formula for making slope corrections is based on the Pythagorean theorem, which gives us:

$$L^2 = S^2 - h^2$$
$$L = \sqrt{S^2 - h^2}$$

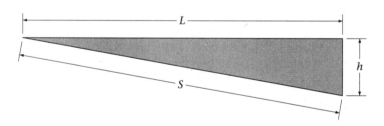

The slope correction (ΔS) is $S - L$ or $S - \sqrt{S^2 - h^2}$.

Because corrections need to be made in the field during chaining, an accurate, simple-to-compute approximation of this

Go Figure!

formula is used. Beginning with a variation of the Pythagorean theorem, we have

$$S^2 = L^2 + h^2$$
$$\Delta S = S - L$$

or:

$$L = S - \Delta S$$

Then, substituting for L in the original equation, we get

$$S^2 = (S - \Delta S)^2 + h^2$$

or:

$$S^2 = S^2 - 2S\Delta S + \Delta S^2 + h^2$$
$$2S\Delta S = \Delta S^2 + h^2$$

To simplify, we assume that ΔS^2 is close to zero. Therefore, we can derive this formula for slope correction:

$$\Delta S = \frac{h^2}{2S}$$

This is a much simpler way to approximate corrections and can be used on most slopes that are not extremely steep.

The following table compares the accuracy of using the approximation formula in calculating slope corrections to using the Pythagorean theorem. In each case, $S = 100$ feet. Even for a 16% slope, the percentage of error is less than 1%.

Accuracy of Approximation Formula Versus Pythagorean Theorem

Percentage of Slope ($h/100$)	ΔS Using Formula	ΔS Using Pythagorean Theorem	Percentage of Error
4	0.080	0.08003	0.04
6	0.180	0.18016	0.09
8	0.320	0.32051	0.16
10	0.500	0.50126	0.25
12	0.720	0.72261	0.36
14	0.980	0.98485	0.49
16	1.280	1.28830	0.65

Go Figure!

Just Between Us Cities
The Gravity Law in Sociology

Have you ever wondered how telecommunication companies estimate the line capacities that will be needed among various cities? They begin with an approximation. It has been shown that the average number of phone calls between two cities in a day (N) is directly proportional to the populations of those cities (P_1 and P_2) and inversely proportional to the square of the distance (d) between the cities:

$$N = \frac{kP_1P_2}{d^2}$$

P_1 and P_2 are expressed in thousands, d is distance in miles, and the constant $k = 400$.

Example: The population of the Minneapolis–St. Paul metropolitan area is 2,538,834, and the population of the Cincinnati–Hamilton metropolitan area is 1,817,571. The distance between the two is 108 miles. We can estimate the average number of telephone calls per day between the two areas using our formula:

$$N = \frac{400 \times 2{,}538.834 \times 1{,}817.571}{108^2} = 158{,}248 \text{ calls per day}$$

These calculations are even simpler with the aid of the pocket calculator:

Press $\boxed{1}\,\boxed{0}\,\boxed{8}\,\boxed{x^2}\,\boxed{1/x}\,\boxed{\times}\,\boxed{4}\,\boxed{0}\,\boxed{0}\,\boxed{\times}\,\boxed{2}\,\boxed{5}\,\boxed{3}\,\boxed{8}\,\boxed{.}\,\boxed{8}\,\boxed{3}\,\boxed{4}$
            $\boxed{\times}\,\boxed{1}\,\boxed{8}\,\boxed{1}\,\boxed{7}\,\boxed{.}\,\boxed{5}\,\boxed{7}\,\boxed{1}\,\boxed{=}$

Display   158,248.0

Example: Let's try two areas at opposite ends of the country. Orlando, Florida, has a population of 1,224,852; the Seattle–Tacoma–Bremerton area has 2,970,320 inhabitants. The distance between the two is 3,403 miles. Therefore:

$$N = \frac{400 \times 1,224.852 \times 2,970.320}{3,403^2} = 126 \text{ calls per day}$$

So, the next time you get a message telling you all circuits are busy, you can estimate how many other callers you're competing with for those precious ten-cent minutes.

Keep in mind that this sociological equation is synonymous with Newton's Law of Gravitation: Two bodies are attracted to each other in direct proportion to the product of their masses, and inversely proportional to the square of the distance between them.

Lottery Fever

It's Everywhere!

Lotteries, originally begun to augment state coffers, have taken the nation by storm. Participation is growing exponentially and gives no indication of slowing. As of 1998, 37 states had lotteries, and 90 million hopefuls bought tickets on a weekly basis. Nor are avid players deterred by the astronomical odds against winning—the hope of that one big jackpot is worth it all!

A variety of games are available, many with themes based on well-known television series or seasonal motifs. Some of the most popular are those based on the "Pick 6" premise, in which players choose six numbers out of 44, 49, or 51. To win, all six numbers (chosen from 1 through 44, 49, or 51) must be drawn during the lottery. Smaller prizes are sometimes awarded for having five or even four numbers, but everyone's eye is on the jackpot, which can often add up to several million dollars!

So, what are your odds of winning? Is there a formula you can use to determine your chances? Of course!

To begin, we use the following formula for the number of combinations of a set of objects (n) chosen a certain number at a time (r). C is the total number of numbers to choose from.

$$nCr = \left(\frac{n}{r} \right)$$

With the parentheses, $\left(\frac{n}{r} \right)$ in this equation is also defined as:

$$nCr = \frac{n!}{r!\,(n-r)!}$$

The n factorial ($n!$) is discussed in detail on pages 34–35. It simply means $n(n-1)(n-2)(n-3)\ldots$ and so on.

Example: Say you want to calculate the odds of winning a Pick 6 game of 49 numbers. Using the formula, you would get:

$$nCr = \frac{49!}{6!(49-6)!}$$

$$= \frac{49 \times 48 \times 47 \times 46 \times 45 \times 44 \times 43!}{6 \times 5 \times 4 \times 3 \times 2 \times 1 \times 43!}$$

$$= \frac{49 \times 48 \times 47 \times 46 \times 45 \times 44}{6 \times 5 \times 4 \times 3 \times 2 \times 1}$$

$$= 13,983,816$$

Thus, the odds of correctly choosing six numbers drawn from 1 through 49 is 13,983,816 to 1.

Most handheld scientific calculators have the *nCr* capability, which simply means finding the total number of combinations of *r* items chosen from a total of *n* different items. For our example, *n* = 49 and *r* = 6. To find the odds for the 6/49 game we do the following:

Press   4   9   nCr   6   =

Display   **13983816**

To find the odds of choosing five numbers correctly in the same game, you simply adapt the formula as shown:

$$nCr = \frac{49!}{5!(49-5)!} = \frac{49 \times 48 \times 47 \times 46 \times 45}{5 \times 4 \times 3 \times 2 \times 1} = 1,906,884$$

The following table shows only some of the games available across the country, the odds of winning the jackpot, and the largest amount awarded to a winner, at the time this book was written.

Lottery Odds and Winning Results

State	Game	Odds	Highest Jackpot ($)
California	6/51	1:18,009,460	118,800,000
Florida	6/49	1:13,983,816	106,500,000
Illinois	6/54	1:25,827,165	69,900,000
Oregon	6/44	1:7,059,052	24,000,000
New York	6/54	1:25,827,165	90,000,000
Pennsylvania	7/74	1:1,799,579,064	115,000,000
20 states plus Washington, D.C.	5/49 plus 1/42	1:80,089,128	195,000,000

Variations on a Theme

Lotteries are so popular that most states have mini-lotteries throughout the week, where participants choose fewer numbers (at lesser odds). For example, the "Pick 3" game, in which players choose three numbers out of ten, is common.

A straight Pick 3 game requires that you choose the correct three numbers in the correct order. To find your odds of winning, you need to modify the premise of the original formula to the number of combinations of a set of objects (n) chosen a certain number at a time (r), with no repeats:

$$n^r$$

For the straight Pick 3 game, $n = 10$ and $r = 3$, so there are 10^3 or 1,000 possible combinations (i.e., the odds of winning are 1 in 1,000).

In some Pick 3 games, you are allowed to "box" the numbers, meaning they do not have to be drawn in any specific order. For example, in a six-way box, you select a three-digit number with three different digits such as 123. You then win if the numbers come up in any of the following combinations: 123, 132, 231, 213, 321, or 312.

Calculating the number of combinations mathematically, you have $3! = 3 \times 2 \times 1 = 6$. Using the formula, you can find your odds of winning:

$$\frac{n^r}{3!} = \frac{10^3}{3 \times 2 \times 1} = 166.67$$

In a Pick 3 three-way box, you select a three-digit number in which two of the digits are the same (they may still appear in any order). If you choose 113, you will win if 113, 131, or 311 is chosen. The number of possible combinations in this game is:

$$\frac{3!}{2} = \frac{3 \times 2 \times 1}{2} = 3$$

Your odds of winning, therefore, are:

$$\frac{n^r}{\left(\frac{3!}{2}\right)} = \frac{10^3}{\left(\frac{3 \times 2 \times 1}{2}\right)} = \frac{1,000}{3} = 333.333$$

Other games that are found in several states are the Lotto Pick 4 with all numbers in the correct order, Lotto Pick 4 with three digits the same, and match 5 out of 39. The number of possible combinations for each and your odds of winning are as follows:

Go Figure!

Lotto Pick 4 (with all numbers in correct order)
Number of combinations:

$$n^r = 10^4 = 10,000$$

Odds of winning are 10,000 to 1.

Lotto Pick 4 (with three digits the same in any order)
Number of combinations:

$$n^r = 10^4 \div \frac{4!}{3!} = 2,500$$

Odds of winning are 2,500 to 1.

Pick 5 (match 5 out of 39)
Number of combinations:

$$nCr = \frac{39!}{5!(39-5)!} = \frac{39 \times 38 \times 37 \times 36 \times 35 \times 34!}{5 \times 4 \times 3 \times 2 \times 1 \times 34!} = 575,757$$

Odds of winning are 575,757 to 1.
Before you risk your money, know the odds. Good luck!

They're Off!
For Thoroughbred Horse Racing Fans

Horse racing, often called "The Sport of Kings," is one of the most popular sports worldwide, drawing millions of patrons year-round. In 1993 alone, approximately 21 billion dollars was wagered at the top forty-seven racetracks in the United States. Over a period of 2,632 days, that equates to 2.4 million dollars per day—and doesn't include the bets placed at the other hundred or so tracks across the nation. Big business, wouldn't you say?

How Fast Can They Go?

Several aspects of horse racing are interesting from a math buff's perspective. Perhaps the most obvious is estimating how long it takes a horse in excellent condition to run a given length on a "fast" (dry) track. For this, we can use the following formula:

$$T = 12.8L - 8.4$$

where T is the time in seconds, and L is the number of furlongs (eighths of a mile) assigned to a particular race. For example, in an 8-furlong race (1 mile), the time would be $12.8 \times 8 - 8.4 = 94$ seconds, or 1 minute, 34 seconds.

When the time calculation results in a fraction of a second, it is rounded to the next fifth of a second. (Running times are traditionally expressed in fifths of a second.)

The table shows the fastest times on record for races of different lengths on twenty U.S. tracks. The estimated time for each race, calculated with our formula, is also shown. As you can see, the formula's accuracy is high. Note that for the relatively short 5-furlong race, $T = 12.8L - 7.2 = 56.8$, or 1.20 seconds more than the basic formula time of $0{:}55^3$, or $55\frac{3}{5}$ seconds.

Comparative Track Records (1994) for Distances Most Frequently Raced at Selected Courses

	5 furlongs	6 furlongs	7 furlongs	8 furlongs	$1\frac{1}{16}$ miles	$1\frac{1}{8}$ miles	$1\frac{1}{4}$ miles
Formula times	$0:55^3$	$1:08^2$	$1:21^1$	$1:34$	$1:40^2$	$1:46^4$	$1:59^3$
Aqueduct	0:57	1:08	1:20	$1:32^2$	—	1:47	1:59
Arlington	$0:57^1$	1:08	1:20	$1:32^3$	1:41	1:46	$1:59^4$
Atlantic City	0:56	1:08	$1:20^2$	—	1:41	$1:46^1$	$2:01^4$
Bay Meadows	$0:56^4$	$1:07^1$	—	$1:33^4$	$1:38^2$	1:46	$2:00^2$
Belmont Park*	0:56	$1:07^4$	$1:20^4$	$1:32^4$	$1:40^2$	$1:45^2$	1:58
Bowie	0:58	1:08	1:20	1:39	1:40	$1:48^4$	$2:03^2$
Calder	$0:58^2$	$1:09^3$	1:23	$1:37^3$	$1:43^4$	1:50	$2:05^1$
Churchill Downs*	$0:57^4$	$1:08^1$	$1:21^4$	$1:33^4$	$1:41^3$	$1:48^2$	1:59
Del Mar	$0:56^2$	$1:07^4$	1:20	1:33	1:40	1:46	$1:59^4$
Garden State	0:56	$1:08^2$	—	1:35	$1:41^3$	$1:45^4$	2:00
Gulf Stream	0:57	$1:07^4$	$1:20^3$	—	$1:40^1$	$1:46^2$	1:59

continued

	5 furlongs	6 furlongs	7 furlongs	8 furlongs	$1\frac{1}{16}$ miles	$1\frac{1}{8}$ miles	$1\frac{1}{4}$ miles
Formula times	$0:55^3$	$1:08^2$	$1:21^1$	1:34	$1:40^2$	$1:46^4$	$1:59^3$
Hileah	0:57	1:08	$1:20^3$	$1:36^3$	$1:40^2$	$1:46^4$	$1:58^3$
Hollywood Park	0:56	1:08	$1:20^4$	$1:32^3$	1:40	$1:46^4$	$1:58^3$
Meadowlands	$0:56^2$	1:08	—	1:35	$1:40^3$	$1:46^2$	$1:58^4$
Monmouth	$0:56^1$	$1:07^4$	—	$1:33^4$	1:41	$1:46^4$	$2:00^2$
Penn National	$0:56^4$	$1:08^4$	—	—	$1:41^1$	$1:49^4$	$2:03^3$
Pimlico*	$0:56^4$	1:09	—	—	$1:40^4$	$1:47^1$	$2:01^4$
Santa Anita	0:58	1:07	1:20	$1:33^2$	1:39	$1:45^4$	$1:57^4$
Saratoga	$0:56^4$	1:09	1:20	$1:34^3$	—	1:47	2:00
Suffolk Downs	0:57	1:08	1:19	1:35	$1:41^4$	$1:47^3$	2:01

*Triple Crown Courses. The Triple Crown record at Pimlico is $1:53^2$ for $1\frac{3}{16}$ miles ($L = 9.5$).
Superscript numbers represent fifths of a second. Statistics for these track records were made available by the Thoroughbred Racing Association (1994).

Remember, when making your own calculations for a "split furlong" race, say $1\frac{3}{16}$ miles, the length is converted to furlongs and the resulting number is a fraction (in this case, 9.5 furlongs).

The average speed in miles per hour is given by:

$$\frac{450L}{12.8L - 8.4}$$

Go Figure!

For $L = 8$ furlongs, we compute the speed as follows:

$$\text{average speed} = \frac{450 \times 8}{12.8 \times 8 - 8.4}$$

$$= 38.3 \text{ miles per hour}$$

Betting

Many tracks feature races offering the opportunity for *combination,* or *ordered,* betting, such as the exacta, quinella, and trifecta. In each case, the player selects two or more horses—the more horses, the greater the number of combinations and the more complex the selection process. Some racing enthusiasts consider combination betting too complicated and avoid potential long shots that could yield big returns.

You can use the following methods to calculate the odds for a variety of specialty races.

Exacta
The object of the *exacta* (also called the *perfecta*) is to pick the first two horses in the correct finish order. The formula for computing the odds of winning an exacta is

$$nPr = \frac{n!}{(n - r)!}$$

where n is the number of horses in the field, and r is the number of horses you need to choose.

Example: If you're playing the exacta for a nine-horse race, $n = 9$ and $r = 2$. Thus:

$$nPr = \frac{n!}{(n - r)!} = \frac{9!}{(9 - 2)!} = \frac{9 \times 8 \times 7!}{7!} = 72$$

Your chances of winning are 72 to 1.

Most handheld scientific calculators have *nPr* capability, which simply means finding the number of permutations of *r* items chosen from a total of *n* different items arranged *in order*. For our example, the operational sequence is:

Press   [9] [INV] [nPr] [2] [=]
Display   **72**

If you want to reduce the odds, you can bet both combinations of the two horses (called *boxing*). Of course, this also doubles the cost of your bet. The following formula is used for calculating the odds on a boxed exacta:

$$nCr = \frac{n!}{r!(n-r)!}$$

Using our earlier example of choosing two horses from a nine-horse field, we have:

$$nCr = \frac{n!}{r!(n-r)!} = \frac{9 \times 8 \times 7!}{2! \times 7!} = 36$$

The odds have obviously been halved.

Most handheld scientific calculators also have the *nCr* capability (shown for the lottery on page 42), which means it has the ability to find the total number of combinations of *r* items chosen from a total of *n* different items. For our example, the operational sequence is:

Press   [9] [nCr] [2] [=]
Display   **36**

Quinella
The object of the *quinella* is to choose the first two horses, which can finish in *either* position. This is the same as *boxing* the exacta, and the odds are calculated using the same formula.

Trifecta
The object of the *trifecta* (also called the *triple*) is to pick the first three horses in exact order of finish. We can use the same formula as for the exacta, plugging in different numbers. For example, choosing three horses from a nine-horse field comes out to:

$$nPr = \frac{9!}{(9-3)!} = \frac{9 \times 8 \times 7 \times 6!}{6!} = 504$$

The trifecta can be boxed just as the exacta can. There are 3! combinations for boxing three horses (e.g., 123, 132, 231, 213, 321, and 312), the formula being:

$$\frac{nPr}{r!} = nCr = \frac{n!}{r!(n-r)!}$$

Using the same nine-horse field as before, the odds for a boxed trifecta are computed as

$$\frac{n!}{r!(n-r)!} = \frac{9!}{3!(9-3)!} = \frac{9 \times 8 \times 7 \times 6!}{3 \times 2 \times 1 \times 6!} = 84$$

or one-sixth the odds for a straight trifecta.

Example: Let's calculate the odds for both an exacta and a trifecta with a 14-horse field. For a straight exacta:

$$nPr = \frac{n!}{(n-r)!} = \frac{14!}{(14-2)!} = \frac{14 \times 13 \times 12!}{12!} = 182$$

For a boxed exacta:

$$nCr = \frac{n!}{r!(n-r)!} = \frac{14 \times 13 \times 12!}{2 \times 1 \times 12!} = 91$$

For a straight trifecta:

$$nPr = \frac{n!}{(n-r)!} = \frac{14 \times 13 \times 12 \times 11!}{11!} = 2,184$$

For a boxed trifecta:

$$nCr = \frac{n!}{r!(n-r)!} = \frac{14 \times 13 \times 12 \times 11!}{3 \times 2 \times 1 \times 11!} = 364$$

Tote-Board Odds

The odds displayed on the tote board before each race represent the payoff odds on each horse for a two-dollar bet or multiples of a two-dollar bet. The amount of the payoff is based on the odds shown just before the betting windows close. Regardless of the odds, the formula for the payoff for a two-dollar bet is:

$$2 \times \text{odds} + \$2$$

For example, let's assume that the tote board shows the odds for a given horse to be 4 to 1. If the horse wins, the payoff will be $2 \times 4 + 2 = 8 + 2 = \$10$.

Payoffs based on fractional odds, such as 5 to 2, can be computed the same way: $2 \times 5/2 + 2 = \$7$.

When the odds are less than 1, the same rule applies. If the odds are 3 to 5, the payoff on a two-dollar bet will be $2 \times 3/5 + 2 = \$3.20$.

You can also reverse the process to calculate the odds for a given payoff. Say you collect eleven dollars for a two-dollar bet. The odds would have been $(11 - 2)/2 = 4\frac{1}{2}$. Converting to the proper fraction form, we get 9/2, or 9 to 2 odds.

So much for thoroughbred horse racing. Next time you go to the track, you'll be better armed. Good luck!

Want to Know the Temperature?

Listen to the Crickets

It is a scientific fact that crickets chirp faster as the temperature rises, as illustrated in the graph. Thus, you can calculate the temperature by counting the number of times a cricket chirps per minute and applying the following formula

$$T = 0.3n + 40$$

where T is the temperature in degrees Fahrenheit and n is the number of cricket chirps per minute.

For example, if you are sitting on your porch and count 50 chirps per minute, the temperature is $T = 0.3 \times 50 + 40 = 55°F$.

Note: The formula seems to indicate that crickets don't chirp at temperatures below 40°F, so it is best to save this method for warmer weather.

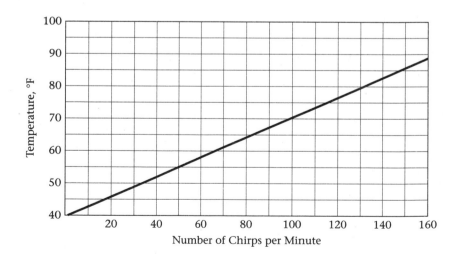

You Can't Prove It by Me!

Goldbach's Conjecture

Christian Goldbach (1690–1760), a Prussian mathematician, said that someone who thinks a statement is true but can't prove it's true may advance that statement as a *conjecture*. For instance, no one has ever found a number greater than 2 that could not be expressed as the sum of two prime numbers. Goldbach said that no such number exists, although no one has been able to prove that this conjecture is true.

Conjecture and Prime Numbers

A prime number is a number that is divisible only by itself and 1, such as 2, 3, 5, 7, 13, and so on. The largest confirmed prime number to date, $2^{1,398,269} - 1$, was found in 1996 by the team of Joel Armengand of Paris and George Woltman of Orlando, Florida. In terms of magnitude, it is 420,921 digits.

Looking at the early prime numbers (excluding the number 2), we see that the sum of any two prime numbers is always an even number. Goldbach made a further conjecture that this pattern holds true for the sum of any two prime numbers. To date, no general formula has been found that gives the number (distribution) of prime numbers below a given integer, so this conjecture also remains to be proven.

Eratosthenes (ca. 276 B.C.–ca. 194 B.C.), the Greek astronomer and mathematician, did, however, discover a "sieve" or systematic way of identifying prime numbers. Using a table of all integers, like the one shown here, cross out every second integer beginning with but excluding 2. Next, cross out every third integer beginning with but excluding 3. Return to the first number after 3 that hasn't been crossed out (5) and cross out every fifth integer beginning with but excluding 5. Continue this pattern through the table. The prime numbers are all those that have not been crossed out.

1	2	3	X̶4̶	5	X̶6̶	7	X̶8̶	X̶9̶	X̶10̶
11	X̶12̶	13	X̶14̶	X̶15̶	X̶16̶	17	X̶18̶	19	X̶20̶
X̶21̶	X̶22̶	23	X̶24̶	X̶25̶	X̶26̶	X̶27̶	X̶28̶	29	X̶30̶
31	X̶32̶	X̶33̶	X̶34̶	X̶35̶	X̶36̶	37	X̶38̶	X̶39̶	X̶40̶
41	X̶42̶	43	X̶44̶	X̶45̶	X̶46̶	47	X̶48̶	X̶49̶	X̶50̶

Go Figure!

Isn't That Stretching It a Bit?

Increasing the Earth's Circumference

The earth's circumference is determined by the formula $C = 2\pi r$. If its radius increases by 1 foot, or one part in 2.092×10^7, how much does the circumference increase?

Given $C = 2\pi r$, we use differential calculus to get the following

$$dC = 2\pi \times dr$$

where dC is the differential increase in circumference, and dr is the differential increase in radius.

If the earth's radius increases 1 foot, we use the formula to get:

$$dC = 2\pi \times 1 \text{ foot}$$
$$= 6.28 \text{ feet}$$

So, imagine a string stretched snugly all the way around the earth. A second string stretched around the earth, but 1 foot above its surface, would need to be only 6.28 feet longer than the first.

When You Grow Up

Predicting Your Child's Height

Parents dream, wonder, and worry about many aspects of their children's future. No formula can tell whether your son or daughter will be president, but there are ways to estimate what height he or she will probably attain as an adult.

The following cubic equations are used with your child's present age (x) and height in inches (h); H represents adult height in inches. The first equation is for girls:

$$H_g = \frac{h}{0.00028x^3 - 0.0071x^2 + 0.0926x + 0.3524}$$

Example: Let's say Gail is 6 years old and is 48 inches tall. How tall will she grow to be? Using our equation, we find that:

$$H_g = \frac{48}{0.00028 \times 6^3 - 0.0071 \times 6^2 + 0.0926 \times 6 + 0.3524}$$

$$= \frac{48}{0.0605 - 0.2556 + 0.5556 + 0.3524}$$

$$= \frac{48}{0.7129} = 67.3 \text{ inches}$$

Thus, Gail will be approximately 5 feet, $7\frac{1}{2}$ inches tall.

Example: To figure out the future height of Gail's baby brother Joe, we would use the following equation:

$$H_b = \frac{h}{0.00011x^3 - 0.0032x^2 + 0.0604x + 0.3796}$$

Go Figure!

Joe is 3 years old and 39 inches tall, so we use the equation to get:

$$H_b = \frac{39}{0.00011 \times 3^3 - 0.0032 \times 3^2 + 0.0604 \times 3 + 0.3796}$$

$$= \frac{39}{0.0030 - 0.0288 + 0.1812 + 0.3796}$$

$$= \frac{39}{0.5350} = 72.9 \text{ inches}$$

Joe's final height will be around 6 feet, $\frac{7}{8}$ inch—a long way from 39 inches!

The same solutions can be found by using the chart on page 60. In Gail's case, we read upward from 6 years until we reach the intersection of the curve for girls. Reading left to the present/future height, we get 0.713. We then divide Gail's present height (48 inches) by 0.713 and get 67.3 inches, her predicted height.

It can be very interesting to record the results of your calculations on a quarterly basis* using the chart.

---

*For fractions of a year in decimal form, see Appendix F.

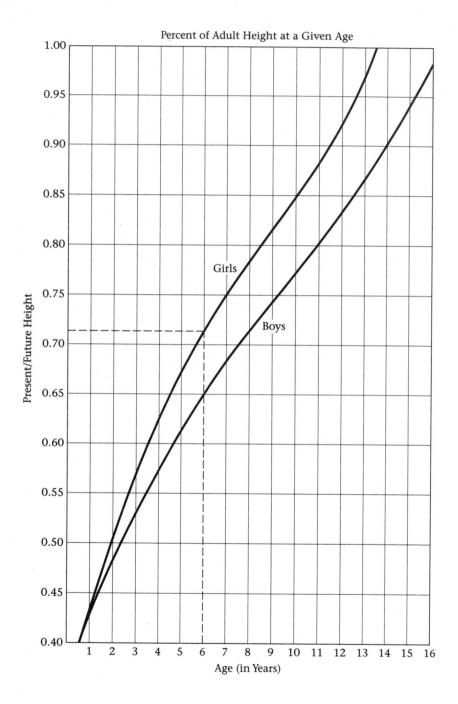

Percent of Adult Height at a Given Age

Go Figure!

It's Not Easy, but It's Fun!

Square Roots, Cube Roots, and nth Roots

In the days before computers and handheld calculators, many mathematical equations took a great deal of time and effort to solve—especially those involving square roots, cube roots, and nth roots. Of course, there were logarithms, but the dedicated mathematician knew there just had to be some other way to do it, some formula to make the whole thing easier.

The root formulas discussed here may seem a bit archaic to the present-day math scholar, but imagine how new and exciting they were to those beleaguered mathematicians who first discovered them. You will find that all of them are still valid in approximating roots of various numbers.

Square Roots

To approximate the square root of a number (n), we first choose a random approximation (x_1), then a better approximation (x_2), and use the following formula:

$$x_2 = \frac{1}{2}\left(x_1 + \frac{n}{x_1}\right)$$

For example, say that $n = 3$ and $x_1 = 1.7$. Using our formula, we get:

$$x_2 = \frac{1}{2}\left(1.7 + \frac{3}{1.7}\right) = 1.7324$$

$$x_3 = \frac{1}{2}\left(1.7324 + \frac{3}{1.7324}\right) = 1.7320508$$

This approximation equals the actual value of $\sqrt{3}$.

Cube Roots

The following formula is used to approximate the cube root of a number (n):

$$n = a^3 + b \cong \left(a + \frac{b}{3a^2} \right)^3$$

In this equation, a^3 represents a perfect cube as close to n as possible. To find b, we plug our numbers into the following equation: $b = n - a^3$.

Example: let's try to approximate $\sqrt[3]{131}$. For a, we choose the number 5, because $5^3 = 125$, which is close to 131. We find b by subtracting: $b = 131 - 125 = 6$.

Using our formula, we find that:

$$n = \left(a + \frac{b}{3a^2} \right)^3 = \left(5 + \frac{6}{3 \times 5^2} \right)^3 = \left(5 + \frac{6}{75} \right)^3 = (5 + 0.08)^3$$

$$\sqrt[3]{n} = \sqrt[3]{131} \cong 5.08$$

Note: This seems unnecessarily complex today, when logarithms can make the process so simple. From $a^n = x$, we get $n \log a = \log x$. In this example, $\frac{1}{3} \times 2.1172713 = \log x$, or $x = 5.0787531$.

*n*th Roots

All real numbers have n number of nth roots (one first root, two square roots, three cube roots, and so on). This is the most interesting formula of all, because it yields an exact solution whether integer solutions exist or not. To approximate the nth root of a number (N), we use the following formula

$$\sqrt[n]{N} = \frac{\left(\dfrac{a^n (n - 1) + N}{n} \right)}{a^{n-1}}$$

where a is our first approximation.

Example: Let's try to find $\sqrt[3]{29}$. We know that 3 is the perfect cube root of 27 (which is close to 29), so we will use $a = 3$.

$$\sqrt[3]{29} \cong \frac{\left(\dfrac{3^3(3-1)+29}{3}\right)}{3^2} = \frac{\left(\dfrac{27(2)+29}{3}\right)}{9} = 3.07407$$

Using this solution, we try a second iteration with $a = 3.074$:

$$\sqrt[3]{29} \cong \frac{\left(\dfrac{3.074^3(2)+29}{3}\right)}{3.074^2} = 3.072318$$

Now, let's prove that the "approximation" can yield an exact solution. If a is an exact solution, then $N = a^n$, so:

$$\sqrt[n]{N} = \frac{\left(\dfrac{a^n(n-1)+a^n}{n}\right)}{a^{n-1}} = \frac{\left(\dfrac{a^n \times n - a^n + a^n}{n}\right)}{a^{n-1}} = \frac{a^n}{a^{n-1}} = a$$

Thus, a is the perfect nth root of N so that $a^n = N$.

How Many of Us Will There Be?
Population Growth in the United States

Since 1960, the first year that both Alaska and Hawaii were included in the national census, the population of the United States has been growing according to the following formula

$$P_x = 0.0036x^{1.77} + 2.25x + 179.3$$

where P is the population in millions, x is the number of years since 1960, and the constant 179.3 is the approximate population in millions for 1960.

Example: Using 1990, the last general census year, $x = 30$ (1990–1960), we get:

$$P_{30} = 0.0036 \times 30^{1.77} + 2.25 \times 30 + 179.3$$
$$= 1.5 + 67.5 + 179.3 = 248.3 \text{ million}$$

This can also be accomplished using a scientific calculator:

Press $\boxed{3}\boxed{0}\boxed{y^x}\boxed{1}\boxed{.}\boxed{7}\boxed{7}\boxed{=}\boxed{\times}\boxed{0}\boxed{.}\boxed{0}\boxed{0}\boxed{3}\boxed{6}$
$\boxed{+}\boxed{2}\boxed{.}\boxed{2}\boxed{5}\boxed{\times}\boxed{3}\boxed{0}\boxed{+}\boxed{1}\boxed{7}\boxed{9}\boxed{.}\boxed{3}\boxed{=}$

Display **248.3**

Example: The same formula can be used to estimate future population. If we wish to project the U.S. population for 2050 ($x = 90$), we get:

$$P_{90} = 0.0036 \times 90^{1.77} + 2.25 \times 90 + 179.3$$
$$= 10.3 + 202.5 + 179.3 = 392.1 \text{ million}$$

Go Figure!

This agrees with the U.S. Bureau of the Census's "middle series" projection of 392,131,000 for 2050, based on an average life expectancy of 82.6 years and a net annual immigration rate of 880,000.

There's a Sucker Born Every Minute
The Pyramid Scheme

Outlawed in most states, the Pyramid is a scam that resurfaces periodically (and aggressively) in various parts of the United States. The premise is for the con artist to convince one person to invest a certain amount of money and get eight of his or her friends to do the same. Then each of these friends must convince eight more people to contribute the same amount of money, and so on.

One of the most highly publicized occurrences of this scheme was uncovered in Washington, D.C., in 1994. Each "investor" was required to contribute $1,500 and recruit eight other players, collecting an additional $12,000. The following chart shows the number of "investors" required to sustain the Pyramid as the process repeats itself and the amount of money collected if the chain remains unbroken.

1 needs $12,000 from
8 who need $96,000 from
64 who need $768,000 from
512 who need $6,144,000 from
4,096 who need $49,152,000 from
32,768 who need $393,216,000 from
262,144 who need $3,145,728,000 from
2,097,152 who need $25,165,824,000 from
16,777,216 who need $201,326,592,000 from
134,217,728 who need $1,610,612,736,000 from
1,073,741,824 who need $12,884,901,888,000 from
8,589,934,592, which is greater than the earth's total population

Figures courtesy of the District of Columbia Department of Consumer and Regulatory Affairs.

Go Figure!

The pattern is simple: the number of participants is 8^n and the number of dollars is $1,500 \times 8^n$. For example, at the seventh level of the Pyramid, the number of people needed to keep the scam going is 8^7, or 2,097,152. When this is multiplied by the mandatory \$1,500 investment, \$25,165,824,000 must be collected from 16,777,216 new investors to move the process to the next level.

So, next time you hear of a get-rich-quick plan that seems too good to be true, do a few calculations of your own. It probably is!

Something Doesn't Add Up!
Baseball Arithmetic

There are occasions when conventional mathematics—and arithmetic, in particular—doesn't reflect the actual situation. Take, for example, baseball players' cumulative batting averages. Fans might mistake "cumulative" to mean the sum of each game's batting average, and not the ratio of cumulative hits to cumulative turns at bat.

For example, each player's batting average is calculated from the ratio of the number of hits to the number of times at bat. If a player gets two hits out of the three times he gets up to bat during one game, and three hits out of four times at bat in a second game, we might mistakenly add the fractions $\frac{2}{3}$ and $\frac{3}{4}$. However, the sum of these fractions is $\frac{17}{12}$, or greater than 1! This can't be true. Even if we divide by the number of games, we still get an average of .708. To get the correct cumulative batting average, which in this case is .714, we must resort to "baseball arithmetic." Using the same example, we simply add the numerators and denominators of $\frac{2}{3}$ and $\frac{3}{4}$, giving us $\frac{5}{7}$, or .714. With this system, the batting average is never greater than 1, no matter how many games are played.

If we wanted to find another player's batting average for four games, we might have the fractions $\frac{1}{3}, \frac{2}{5}, \frac{3}{5}$, and $\frac{1}{4}$. Adding the numerators and denominators, we get $\frac{7}{17}$, or .412.

This may seem to violate the rules of arithmetic, but it works for baseball!

It Took Only 350 Years to Prove!
Fermat's Last Theorem

The avid math buff will enjoy this one. Pierre de Fermat (1601–1665), the French mathematician, came up with one of the most famous theorems of all time (it was also the last theorem he made):

> *Equations of the form* $x^n + y^n = z^n$ *have no solution when n is a whole number greater than 2 and when* x, y, *and* z *are positive whole numbers.*

When *n* equals 2, the equation is the Pythagorean theorem, which says that the sum of the squares of the lengths of the two legs of a right triangle equals the square of the hypotenuse. Probably the most popular example of this, known as the 3-4-5 triangle, is $3^2 + 4^2 = 5^2$.

Fermat's last theorem has tantalized mathematicians for more than 350 years. He had scribbled the proof for his theorem in the margin of a book, adding that although the proof was excellent, the space available was inadequate to hold it. Dr. Enrico Bombieri, a scholar at the Institute for Advanced Study, Princeton University, even said, "Everyone has a price. For mathematicians, it's the proof of Fermat's last theorem."

On June 24, 1993, Dr. Andrew Wiles of Princeton University made the stunning announcement that he had indeed discovered this proof.[1] This naturally aroused worldwide attention. Unfortunately, Wiles's proof turned out to have a gap which he was unable to fill due to his reluctance to accept (for many months) Dr. Mathias Flach's method of "sophisticated mechanics." Contemporaries criticized the validity of his evidence.

---

[1]*Notes on Fermat's Last Theorem*, by Alf van der Poorten (John Wiley & Sons, 1996), provides a succinct discussion of Andrew Wiles's proof.

As time went on and he was unable to complete his proof, Wiles faced a dilemma. If he were to invite another well-known mathematician to help him bridge the gap, he would risk the necessity of sharing the credit. However, without help, he stood a good chance of losing the intellectual trophy of having solved the world's most famous mathematical problem. As Dr. Kenneth Ribet, University of California at Berkeley, stated, "A proof that is unfinished is no proof at all."

After careful consideration, Wiles asked Dr. Richard Taylor, a 32-year-old former student and reader with tenure at Cambridge University, to help. Elated and excited about the opportunity, Taylor took a sabbatical from Cambridge and joined Wiles at Princeton in December 1993. After several months, Taylor suggested that they try a powerful new method devised by Dr. Flach, which Wiles had considered previously and discarded.

Going back to Flach's method for the second time, after long hours of deliberation Wiles found that the very thing he had seen as useless was the one thing that completed his proof! The end was in sight, but it needed further development.

He called Taylor, who had returned to England; the two worked feverishly over the next two and a half weeks to prepare a paper that filled the gap in Wiles's proof. By agreement, the final proof was called "The Theorem of Wiles, completed by Taylor and Wiles."

It's All Around Us
The Fibonacci Series

It would be unfair to end the first part of this book without mentioning one of the most interesting phenomena in all of mathematics: the Fibonacci series. Leonardo de Pisa (better known as Fibonacci) published an observation pertaining to the propagation of rabbits in A.D. 1202. This observation has since been found to relate to botany, biology, art, number theory, astronomy, and music.

Each number in the Fibonacci series (0, 1, 1, 2, 3, 5, 8, 13, 21, 34, 55, 89, 144, . . .) is the sum of the two numbers that precede it. The most interesting aspect of this series is the relationship of the ratios of successive numbers, such as 3 to 5, 5 to 8, 8 to 13, and so on. As the ratios increase, they get closer to 0.618034; in terms of n numbers, this relationship is expressed:

$$\lim_{n \to \infty} \frac{n-1}{n} = 0.618034$$
$$= \frac{(\sqrt{5} - 1)}{2}$$

The "golden number" of 0.618034 is denoted by the Greek letter phi (Φ). It can be shown that:

$$1 + \Phi = 1/\Phi$$

We can find Φ algebraically by rewriting this equation as a quadratic equation and then solving it with the quadratic formula:

$$\Phi + \Phi^2 = 1$$
$$\Phi^2 + \Phi - 1 = 0$$
$$\Phi = \frac{-1 \pm \sqrt{1^2 - 4 \times 1 \times -1}}{2 \times 1} = \frac{-1 \pm \sqrt{5}}{2}$$

Therefore, Φ = 1.618034 or 0.618034. Accepting that Φ = 0.618034, we find that $1 + \Phi$ = 1.618034 and:

$$1.618034 = \frac{1}{0.618034}$$

Geometrically, Φ can be shown to be unique in the "golden rectangle," where $R = \Phi + \frac{1}{2} = \frac{\sqrt{5}}{2}$:

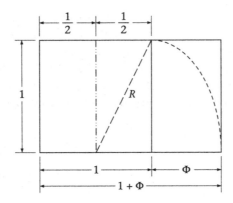

This rectangular relationship can be found in the dimensions of playing cards, index cards, briefcases, billboards, and many architectural designs.

In nature, the Fibonacci series also occurs in plants. For instance, lilies have three petals; buttercups five; marigolds thirteen; asters twenty-one; and daisies thirty-four, fifty-five, or eighty-nine.

Go Figure!

It is interesting to note that the Fibonacci series is not unique to the limiting value of $(n - 1) \div n = 0.618034$. Let's try starting with 2, 5, 7, 12, 19, and so on. Proceeding to the thirteenth and fourteenth terms—898 and 1,453, respectively—we find that the ratio of 898 to 1,453 is 0.618032, which is close to 0.618034. The same relationship holds true for any combination of starting numbers that are less than 10.

Next, let's see how the golden rectangle can be used to construct a five-point star from the unique relationship of Φ and $1 + \Phi$. By using the proper arcs to construct an equilateral triangle with sides equal to 1 and 1.618 – 1, as shown at the bottom of the page, we see that the angles at the apexes are exactly 36 degrees—the same as those at the points of a five-point star. The golden triangle premise works in architecture as well; the ratio of the sum of the areas of the four triangular faces to the area of the base of the Great Pyramid of Giza is $1 + \Phi$.

You can see that the applications for the Fibonacci series are never-ending. There is even a *Fibonacci Quarterly,* founded in 1963, devoted to the series and its related forms.

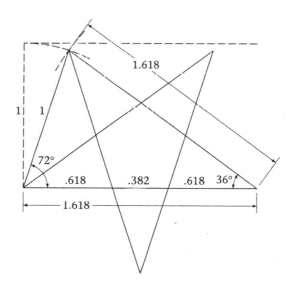

For pure math buffs: What is the value of the *n*th term of the Fibonacci series? Here's a hint

$$a_n = \frac{1}{\sqrt{5}}\left[\left(\frac{1 + \sqrt{5}}{2}\right)^{n-1} - \left(\frac{1 - \sqrt{5}}{2}\right)^{n-1}\right]$$

where $a_1 = 0$.

PART II

Workings of
Time and Space

Solar System Mechanics

There is nothing more intriguing to the math buff than the mechanics of the solar system—a perfectly ordered scheme of nine planets, some with satellites or moons, orbiting around a star. There are voluminous writings available regarding Newtonian physics and its application to the bodies of the solar system. In this book, we will content ourselves with a brief overview (Appendix E gives additional commentary). Formulas defining mathematical relationships are presented in their simplest form; our focus is on practical problems and their solutions.

Isaac Newton (1642–1727) published his *Principia Mathematica* in 1686, twenty years after completing his laws of gravitation. Difficulty had arisen because in 1666 the radius of the earth was thought to be about 13 percent less than it actually was. This error was corrected in 1671, when more reliable data became available, and Newton was able to confirm his calculations.

The *Principia* made an enormous impact. Its three laws of motion and single universal law of gravitation explained every-thing about the motions of celestial bodies. The latter defines the mutual attraction between all masses and particles of matter in the universe, which is, in a sense, one of the best-known physical phenomena. During the eighteenth and nineteenth centuries, gravitational astronomy, based on Newton's laws, attracted many of the leading mathematicians. It was studied so extensively, it seemed that only further numerical refinements were needed to give detailed accounts of the motions of all bodies in the solar system. This view was shattered by Albert Einstein (1879–1955), and the subject is currently in a healthy state of flux.

Until the seventeenth century, the sole recognized evidence of attraction between physical bodies was the gravitational attraction to the surface of the earth. There was only vague speculation that some force emanated from the sun to keep the planets in their orbits. Such a view was expressed by Johannes Kepler (1571–1630), the author of the laws of planetary motion.

It is doubtful that Newton could have developed his law of gravitation without reference to Kepler's work, which included the following three laws:

1. Each planet moves in an elliptical orbit with the sun as a focus.

2. The rays from the sun to the planet sweep out equal areas of the ellipse in equal times.

3. The square of the period of a planet is proportional to the cube of the mean distance between the planet and the sun.

These laws are illustrated in the figure on page 78.

Newton was able to combine Kepler's three laws into one inverse-square law. The problem was to prove that a force, varying in strength as the inverse square of the distance, would require a planet to travel in an ellipse, thereby establishing that the force behaves in a manner consistent with Kepler's laws. Newton not only extracted underlying physical laws from an excess of observational data, he invented a new labor-saving mathematical tool—the calculus—with which to express those relationships.

Newton's law of gravitation states that two particles of matter attract each other with a force that acts as a line joining them. The intensity of this force varies as the product of the masses and inversely as the square of the distance between them. This is represented by

$$F = \frac{Gm_1m_2}{d^2}$$

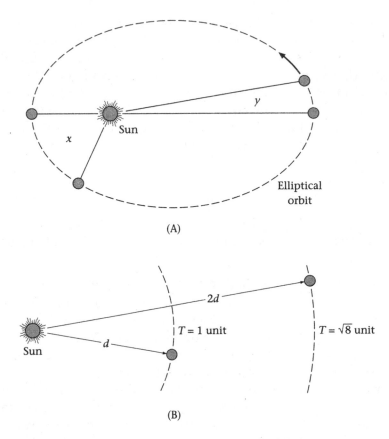

(A)

(B)

According to Johannes Kepler, the orbits of the planets were ellipses with the sun located at one focus (A). The speed of each planet around the sun was such that equal areas such as x and y would be traversed in equal lengths of time by a line connecting the planet with the sun. The revolution period of a planet was shown to be dependent on its distance from the sun. Doubling the distance would increase the period by a factor or $\sqrt{2^3}$, or $\sqrt{8}$, as shown at B.

where F is the gravitational force exerted by two particles, of masses m_1 and m_2, separated by a distance (d), which is generally used for all celestial bodies. G is the absolute constant of gravitation; it is independent of time, place, and the chemical composition of the masses involved.

Go Figure!

Newton verified that the gravitational force between the earth and the moon, necessary to maintain the moon in its orbit, and gravitational attraction at the earth's surface were related to the inverse-square law of force. Let m_e be the mass of the earth and assume it is spherically symmetrical with radius r. The force exerted by the earth on a smaller mass (m_m) such as the moon is then given by

$$F = \frac{Gm_em_m}{r^2}$$

We use the radius of the earth, and not the distance between the earth and the moon, because we are measuring the gravitational force *at the earth's surface* in this local system. However, m_m is insignificant compared to m_e, so F or g (the acceleration of gravity at the earth's surface) is:

$$F = g = \frac{Gm_e}{r^2}$$

For practical purposes, the numerical value of the gravitation constant is:

$$G - 1.07 \times 10^{-9}\,\text{ft}^3/\text{lb sec}$$

All problems and examples in this section use English units of measure; a metric conversion table appears in Appendix A.

The table on page 80 provides data on solar system bodies that will be used later in this section. You may also use the values in solving solar system problems not found in this book.

The topics presented in Part II are intended to whet your appetite for discovering more about how the solar system works. The math in this section is a bit more sophisticated than that in Part I, but a brief review of the material in the appendixes to this book should give you any additional help you need. It's also

Comparative Data for Solar System Bodies

Solar body	Mass (lb)	Radius (mi)	Density (lb/ft³)	Period (yr)	Period (d)	Distance to sun (AUs)	Escape velocity (mi/sec)	Surface gravity (ft/sec²)	Eccentricity
Mercury	7.2819×10^{23}	1,515	338	0.241	87.97	0.387	2.643	12.177	0.2056
Venus	1.07365×10^{25}	3,759	328	0.615	224.70	0.723	6.444	29.163	0.0068
Earth	1.31748×10^{25}	3,963	343	1.000	365.26	1.000	6.952	32.197	0.0167
Mars	1.41559×10^{24}	2,107	246	1.881	686.98	1.524	3.125	12.238	0.0934
Jupiter	4.1866×10^{27}	44,448	77	11.862	4,332.59	5.203	37.005	81.334	0.0485
Saturn	1.25355×10^{27}	37,380	43	29.46	10,759.22	9.539	22.080	34.433	0.0556
Uranus	1.91188×10^{24}	16,079	74	84.01	30,685.4	19.182	13.148	28.383	0.0472
Neptune	2.2528×10^{24}	15,379	104	164.79	60,189.1	30.058	14.613	36.655	0.0086
Pluto	3.294×10^{22}	713	62	248.5	90,766.0	39.526	0.820	2.487	0.2540
Ceres*	2.635×10^{21}	347	43	4.61	1,682.0	2.77	0.332	0.840	0.077
Sun	4.386×10^{30}	4.325×10^{5}	88	$2 \times 10^{8\dagger}$	—	$2.15 \times 10^{9\ddagger}$	383.97	899.94	—

*The largest asteroid in the solar system.
†Period for sun revolving around its galactic center.
‡Radial distance to galactic center (AUs).

Go Figure!

helpful to remember that 5,280 feet = 1 mile and that 5,280 = 5.28×10^3. You'll see this figure in many of the equations found in the rest of this book.

Before continuing, it is a good idea to review the notes on scientific notation and exponentials and radicals in Appendixes B and C. Solar system calculations frequently involve exponents and very large or very small numbers; thus, using scientific notation can be expedient in problem solving. As you work the problems, remember that the solutions are close *approximations;* that is, ellipses have been made circular, radii of planets averaged, solar constants rounded, and so on. However, the main thrust of each solution is still valid.

Just think: when you're done with this section, you will be able to figure the weight of your favorite planet or compute how quickly the pull of gravity vanishes as astronauts leave the earth and how it increases as they near the moon. Have fun!

Calculator Math Made Simple

The Power of Scientific Notation

Solar system problems necessarily involve large numbers, such as distances between the sun and planets and diameters of large solar bodies. Scientific notation, a method for dealing with very large and small numbers, is convenient for handling mathematical problems pertaining to physical measurements in the solar system. Before trying the example problems that follow, review Appendix B, which clearly illustrates the scientific notation process.

The typical problems presented in Part II of *Go Figure!* fall into a limited number of categories, beginning with simple applied calculator math and building to more complex groups of expressions. The following exercises represent the math operations you will need to do to solve solar system problems. Although dimensions used in the problem solutions are identified as multidigit numbers, the calculator math examples are simplified by using single digits—the principles are still the same! You will find that using a calculator simplifies the required operations considerably.

Example: Solve for $(5 \times 10^3)(4 \times 10^3)$.

Press   [5] [EE↓] [3] [X] [4] [EE↓] [3] [=]

Display   2 07

Example: Solve for $(5 \times 10^3)(4 \times 10^3)^2$:

Press   [5] [EE↓] [3] [X] [4] [EE↓] [3] [x^2] [=]

Display   8   10

Example: Solve for $[(5 \times 10^3)(4 \times 10^3)]^2 \times 10^9$:

Press   [5] [EE↓] [3] [X] [4] [EE↓] [3] [=] [x^2] [X] [1] [EE↓] [9] [=]

Display   4   23

Example: Solve for:

$$\frac{[(5 \times 10^3)(4 \times 10^3)]^2 \times 10^9}{\pi(2 \times 10^7)}$$

Press   [2] [EE↓] [7] [X] [π] [=] [STO] [5] [EE↓] [3] [X] [4] [EE↓] [3] [=] [x^2] [X] [1] [EE↓]
                [9] [÷] [RCL] [=]

Display   6.366   15

Example: Solve for:

$$\frac{(2\pi \times 10^8)^{\frac{3}{2}}}{(2 \times 10^{16})^{\frac{1}{2}}}$$

Press   2 EE↓ 1 6 √x̄ STO 2 X π = X 1 EE↓ 8 = yˣ 1 · 5

              = ÷ RCL =

Display   1.11367 05

You should now have little trouble in using a calculator to work the rest of the problems in this section.

Properties of Solar System Bodies

Weight of the Earth

To find the earth's weight, start with the formula for surface gravity:

$$g = \frac{Gm_e}{r^2}$$

From the table on page 80, we know that $g = 32.197$ ft/sec^2 and $r = 3,963$ miles. Reorganizing the formula, we get:

$$m_e = \frac{gr^2}{G}$$

Solve for m_e by substituting for the variables and converting miles to feet:

$$m_e = \frac{32.197 \times [(3.963 \times 10^3) \times (5.28 \times 10^3)]^2 \times 10^9}{1.07}$$

$$= \frac{32.197 \times (437.8406 \times 10^{12}) \times 10^9}{1.07}$$

$$= \frac{1.4097 \times 10^{25}}{1.07} = 1.3175 \times 10^{25} \text{ lb}$$

The operational sequence for solving this problem with a scientific calculator is:

Press   3 · 9 6 3 EEi 3 × 5 · 2 8 EEi 3

        = x^2 × 3 2 · 1 9 7 EEi 9 = ÷ 1 · 0 7 =

Display   1.3175 25

Confirmation of the Earth's Weight

The mass of a planet may be found easily if it has one or more satellites. If d is the mean distance of the satellite's orbit from the planet and T is its period of revolution, expressed respectively in astronomical units (AUs) and siderial years, the mass of a planet (m) is given through Newton's law of gravitation by

$$m = \frac{d^3}{T^2}$$

in terms of the mass of the sun as a unit.

To confirm the earth's mass, we begin by expressing the distance from the earth to the moon (d_m) in AUs, where $1\text{AU} = 9.3 \times 10^6$ miles:

$$d_m = \frac{238{,}218 \text{ mi}}{93{,}000{,}000 \text{ mi}} = \frac{2.38218 \times 10^5}{9.3 \times 10^7} = 2.5615 \times 10^{-3} \text{ AU}$$

$$d_m^3 = 1.680672 \times 10^{-8}$$

$$T = 27 \text{ days, 7 hr, 43 min, 11.6 sec}$$
$$= 27 \text{ days, 7 hr, 43.1933 min}$$
$$= 27 \text{ days, 7.7199 hr} = 27.3217 \text{ d}$$

Now convert the number of days into decimal form, representing the number of years:

$$T = 27.3217 \text{ d} = \frac{27.3217}{365.26} \text{ yr}$$
$$= 0.0748 \text{ yr} = 7.48 \times 10^{-2} \text{ yr}$$

Next, square the period of revolution (T):

$$T^2 = 5.595 \times 10^{-3}$$

Finally, substitute the values into the formula $m_e = \dfrac{d_m^3}{T^2} \times m_s$ (m_s is the mass of the sun, or 4.386×10^{30}):

$$m_e = \frac{1.680672 \times 10^{-8}}{5.595 \times 10^{-3}} \times 4.386 \times 10^{30}$$

$$= 3.00388 \times 10^{-6} \times 4.386 \times 10^{30}$$
$$= 1.3175 \times 10^{25} \text{ lb}$$

Press   [1] [.] [6] [8] [0] [6] [7] [2] [EE↓] [8] [+/-] [÷] [5] [.] [5] [9] [5]

           [EE↓] [3] [+/-] [X] [4] [.] [3] [8] [6] [EE↓] [3] [0] [=]

Display   1.3175 25

Density of the Earth

The formula for density is:

$$\text{density} = \frac{\text{mass}}{\text{volume}}$$

To solve for the earth, we use values for mass and radius from the table on page 80. A sphere's volume equals $\frac{4}{3}\pi r^3$. (Multiply the radius by 5,280 to convert it to feet.)

$$\text{density} = \frac{1.3175 \times 10^{25}}{\frac{4}{3}\pi(3.963 \times 10^3 \times 5.28 \times 10^3)^3} = \frac{1.3175 \times 10^{25}}{\frac{4}{3}\pi(2.092464 \times 10^7)^3}$$

$$= \frac{1.3175 \times 10^{25}}{3.837626 \times 10^{22}} = 3.433 \times 10^2 = 343.3 \text{ lb/ft}^3$$

Press   [4] [÷] [3] [X] [π] [X] [2] [.] [0] [9] [2] [4] [6] [4]

           [EE↓] [7] [y^x] [3] [=] [STO] [1] [.] [3] [1] [7] [5] [EE↓] [2] [5] [÷] [RCL] [=]

Display   343.3

Weight of the Moon

The formula used on page 85 for the weight of the earth is still applicable; only the variables change. Given that the moon's surface gravity (g) is 5.33 ft/sec^2, its radius (r) is 1,080 miles,

Workings of Time and Space

and its mass is m_m, we can reorganize the formula (converting miles to feet) to get:

$$m_m = \frac{gr^2}{G} = \frac{5.33(1.08 \times 10^3 \times 5.28 \times 10^3)^2}{1.07 \times 10^{-9}}$$

$$= \frac{5.33(3.25174 \times 10^{13}) \times 10^9}{1.07}$$

$$= 1.62 \times 10^{23} \text{ lb}$$

Density of the Moon

We again use the formula

$$\text{density} = \frac{\text{mass}}{\text{volume}}$$

and plug in new values for the moon:

$$\text{density} = \frac{1.62 \times 10^{23}}{\frac{4}{3}\pi(1.08 \times 10^3 \times 5.28 \times 10^3)^3}$$

$$= \frac{1.62 \times 10^{23}}{\frac{4}{3}\pi(1.85427 \times 10^{20})}$$

$$= \frac{1.62 \times 10^{23}}{7.767 \times 10^{20}}$$

$$= 209 \text{ lb/ft}^3$$

Weight of Jupiter

From the data for Jupiter in the table on page 80, we find that $g = 81.334$ ft/sec^2 and $r = 44,448$ miles. Following the same procedure used for the weight of the earth, we get:

$$m_j = \frac{81.334(44,448 \times 5.28 \times 10^3)^2}{1.07 \times 10^{-9}}$$

Go Figure!

$$= \frac{81.334(2.3468544 \times 10^8)^2 \times 10^9}{1.07}$$

$$= \frac{81.334(5.507726 \times 10^{16}) \times 10^9}{1.07}$$

$$= \frac{4.47965 \times 10^{27}}{1.07}$$

$$= 4.1866 \times 10^{27} \text{ lb}$$

Density of Jupiter

We know that density equals mass divided by volume. Using values for Jupiter from the table on page 80, we find:

$$\frac{\text{mass}}{\text{volume}} = \frac{4.1866 \times 10^{27}}{\frac{4}{3}\pi(23.468544 \times 10^7)^3}$$

$$= \frac{4.1866 \times 10^{27}}{5.4144 \times 10^{25}}$$

$$= 77 \text{ lb/ft}^3$$

Weight of Pluto

Published physical statistics for our most distant planet, Pluto, vary greatly. Perhaps it is because it's so far away that astronomers have difficulty in obtaining accurate data. Imagine—it takes just less than 250 years for Pluto to make one revolution around the sun! Our solution here doesn't guarantee any more accuracy than those reported to date, but it does confirm that Newton's laws play no favorites when it comes to their validity for distant bodies in our solar system. We will use Newton's formula as demonstrated for confirming the earth's weight (page 86).

Given that $T = 6.39$ days for Pluto's moon Charon to make one revolution,

$$T = \frac{6.39}{365.26} = 0.0175 \text{ yr}$$

$$T^2 = (1.75 \times 10^{-2})^2 = 3.0625 \times 10^{-4}$$

$$d_C = \frac{12{,}276 \text{ mi}}{93{,}000{,}000 \text{ mi}} = 0.000132 = 1.32 \times 10^{-4} \text{ AU}$$

$$d_C^3 = 2.300 \times 10^{-12}$$

$$m_p = \frac{d_C^3}{T^2} \times m_s = \frac{2.300 \times 10^{-12}}{3.0625 \times 10^{-4}} \times 4.386 \times 10^{30} = 3.294 \times 10^{22} \text{ lb}$$

Surface Acceleration of the Sun

From the formula on page 85, we know that:

$$g = \frac{Gm_s}{r^2}$$

Given that $m_s = 4.386 \times 10^{30}$ pounds and $r = 4.325 \times 10^5$ miles, substitute, converting miles to feet:

$$g_s = \frac{1.07 \times 10^{-9} \times 4.386 \times 10^{30}}{(4.325 \times 10^5 \times 5.28 \times 10^3)^2}$$
$$= 899.94 \text{ ft/sec}^2$$

Press   [4] [.] [3] [2] [5] [EE⁺] [5] [X] [5] [.] [2] [8] [EE⁺] [3] [=] [x²] [X]

           [1] [EE⁺] [9] [=] [STO] [4] [.] [3] [8] [6] [EE⁺] [3] [0] [X]

           [1] [.] [0] [7] [=] [÷] [RCL] [=]

Display   899.94

Surface Acceleration of the Moon

Given that $m_m = 1.62 \times 10^{23}$ pounds and $r = 1{,}080$ miles, we solve for the formula

$$g = \frac{Gm_m}{r^2}$$

$$g_m = \frac{1.07 \times 10^{-9} \times 1.62 \times 10^{23}}{(1.08 \times 10^3 \times 5.28 \times 10^3)^2}$$
$$= 5.33 \text{ ft/sec}^2$$

Go Figure!

Escape Velocity

Envision a projectile being launched from the center of the earth. To counteract the force of gravity and move into infinite space by the time it reaches the earth's surface, the object must achieve a velocity of $v = \sqrt{2gr}$, where g is the earth's surface gravity, and r is equal to the earth's radius. Remembering that 5,280 feet equals one mile, solving for v, we get:

$$v = \frac{\sqrt{2 \times 32.197 \times 3.963 \times 10^3 \times 5.28 \times 10^3}}{5,280} \text{ mi/sec}$$

$$= \frac{\sqrt{1347.42 \times 10^6}}{5.28 \times 10^3}$$

$$= 6.95 \text{ mi/sec}$$

You can also use a handheld calculator to solve this:

Press   3 · 9 6 3 X 5 · 2 8 EE⁺ 6 X 2

          X 3 2 · 1 9 7 = √x̄ ÷ 5 2 8 0 =

Display   6.95

If you want to continue making calculations, you may find it helpful to use the table on page 92.

Properties of Solar System Bodies

Solar System Body	Mass (m) (lb)	$G_m = 1.07 \times 10^{-9} \times m$	Radius (mi)	Radius(r) (ft)	r^2	$Gm/r^2 = g$	$\sqrt{2gr}$	Escape Velocity (v) ($\div\ 5.28 \times 10^3$)
Venus	10.7365×10^{24}	1.1488×10^{16}	3759	1.9847×10^{7}	3.939×10^{14}	29.165	3.4025×10^{4}	6.444
Jupiter	4.1866×10^{27}	4.4797×10^{18}	44448	2.3469×10^{8}	5.508×10^{16}	81.331	1.9538×10^{5}	37.005
Sun	4.386×10^{30}	4.6930×10^{21}	4.325×10^{5}	2.2836×10^{9}	5.215×10^{18}	899.91	2.0273×10^{6}	383.97
Moon	1.620×10^{22}	1.7334×10^{14}	1080	5.7024×10^{6}	3.252×10^{13}	5.33	7.797×10^{3}	1.48
Ceres	2.635×10^{21}	2.8195×10^{12}	347	1.8322×10^{5}	3.357×10^{10}	0.84	1.754×10^{3}	0.33

Go Figure!

Earth-Moon Point of Equilibrium

There is a point at which a body in space will be drawn to neither the earth nor the moon. At what point are the gravitational forces of these two bodies equal?

For this calculation, we will use the following numbers: the earth's radius = 3,963 miles; the moon's radius = 1,080 miles; the earth's surface acceleration = 32.197 feet/second; and the moon's surface acceleration = 5.33 feet/second. We next apply Newton's inverse-square law

$$\left(\frac{1,080}{d_m}\right)^2 \times 5.33 = \left(\frac{3,963}{d_e}\right)^2 \times 32.197$$

where d_m is the distance to the moon from the point of equilibrium; d_e is the distance to the earth from the point of equilibrium; and $d_m + d_e = 238.22 \times 10^3$ miles.

Solving for this, in terms of 10^3 miles, we get:

$$\left(\frac{1.080}{d_m}\right)^2 \times 5.33 = \left(\frac{3.963}{d_e}\right)^2 \times 32.197$$

$$\frac{6.217}{d_m^2} = \frac{505.67}{d_e^2}$$

$$6.217 d_e^2 = 505.67 d_m^2$$

$$2.4934 d_e = 22.4871 d_m$$

Given that $d_m + d_e = 238.22$, in terms of 10^3 miles, we can solve simultaneous equations:

$$d_e = 238.22 - d_m$$

$$2.4934(238.22 - d_m) = 22.4871 d_m$$

$$593.9777 - 2.4934 d_m = 22.4871 d_m$$

$$24.9805 d_m = 593.9777$$

$$d_m = 23.777 \text{ in terms of } 10^3 \text{ miles}$$

Now solve for both variables in terms of miles:

$$d_m = 23.777 \times 10^3 \text{ mi}$$
$$d_e = (238.220 - 23.777) \times 10^3 = 214.443 \times 10^3 \text{ mi}$$

Working backward, we can prove the validity of this answer:

$$\left(\frac{1.080}{23.777}\right)^2 \times 5.33 = \left(\frac{3.963}{214.443}\right)^2 \times 32.197$$

$$0.010996 = 0.010996 \text{ ft/sec}^2 \text{ at point of equilibrium}$$

Synchronous Satellites

One of the most significant advances in recent decades has been the establishment of a system of artificial satellites that remain in fixed positions at a certain height above the earth's equator. These *synchronous satellites* (or *tracking and data relay satellites*) are equipped to relay transmissions throughout the intercontinental communications network.

The height at which the satellites orbit is determined by the fact that the satellites' period of rotation must exactly equal that of the earth. If the periods are equal, the satellites and the earth rotate together, and the satellites remain in fixed positions with respect to the earth's surface.

We have seen that the force of gravitation (F) between two masses (m_1 and m_2) is given by Newton's inverse-square law:

$$F = \frac{Gm_1m_2}{r^2}$$

According to Newton's second law, the centripetal force exerted on mass m_2 must equal the mass of the object times the centripetal acceleration (m_2v^2/r):

$$\frac{Gm_1m_2}{r^2} = \frac{m_2v^2}{r} \quad \text{or} \quad \frac{Gm_1}{r^2} = \frac{v^2}{r}$$

Solving for v^2, we get

$$v^2 = \frac{Gm_1}{r} \quad \text{or} \quad r = \frac{Gm_1}{v^2}$$

If r is the distance from the center of the earth to the satellite of mass m_s, then v is the velocity required to prevent the satellite from falling toward the earth.

We know that $Gm_e = 1.4 \times 10^{16}$, and v is the velocity of the satellite rotating around the earth in twenty-four hours. Converting to feet per second, we get:

$$v = \frac{2\pi r}{24 \times 3{,}600} = \frac{2\pi r \text{ ft}}{8.64 \times 10^4 \text{ sec}}$$

Next, square the equation to find v^2:

$$v^2 = \frac{(2\pi r)^2}{7.46496 \times 10^9} = \frac{39.4784 r^2 \text{ ft}^2}{7.46496 \times 10^9 \text{ sec}^2}$$

Now substitute to solve for r:

$$r = \frac{Gm_1}{v^2} = \frac{(1.4 \times 10^{16}) \times (7.46496 \times 10^9)}{39.4784 r^2}$$

$$r^3 = \frac{(1.4 \times 10^{16}) \times (7.46496 \times 10^9)}{39.4784} = 2.647256 \times 10^{24}$$

$$r = 1.38335 \times 10^8 \text{ ft} \quad \text{Converting to miles,}$$

$$r = 26{,}200 \text{ mi}$$

Subtracting the earth's radius from this number, we get $26{,}200 - 3{,}963 = 22{,}237$. Thus, the satellites should be 22,237 miles above the earth's surface. A satellite orbiting at this height will appear stationary to an observer on earth. Thus, such satellites are often called *geostationary*.

To solve this problem using a calculator, the operational sequence is:

Press   1 · 4 X 7 · 4 6 4 9 6 X 1 EE+ 2 5 ÷
        3 9 · 4 7 8 4 y^x · 3 3 3 3 3 3 =
        ÷ 5 2 8 0 = − 3 9 6 3 =

Display   22237

Go Figure!

Kepler's Third Law

Johannes Kepler (1571–1630) observed that:

The square of the period of revolution of any planet is a constant multiplied by the cube of that planet's distance from the sun.

Thus, $T^2 = Kd^3$ is the basic formula where:

$$K = \frac{4\pi^2}{Gm}$$

Depending on the desired units for T, and expressing d in astronomical units (AU), the following table shows the corresponding K values:

T value	d value (mi)	K value	
Years	$d_{AU} \times 93$	1.243325×10^{-6}	(A)
Days	$d_{AU} \times 93$	0.165863	(B)

Through interaction of the variables in the basic formula, the following relationships are also useful in calculating periods of revolution and distances from the sun to various planets:

$$T_{years} = d_{AU}^{1.5} \qquad\qquad \text{(C)}$$

$$d_{AU} = (T_{years})^{0.6667} \qquad\qquad \text{(D)}$$

$$\frac{T^2_{\text{years}}}{d^3_{\text{AU}}} = 1 \qquad\qquad (E)$$

and:

$$T^2 = Kd^3$$
$$T = \sqrt{Kd^3}$$
$$= \sqrt{(1.243325 \times 10^{-6})(9.539 \times 93)^3}$$

Let's use our formula and the other relationships listed to prove the validity of Kepler's third law for Saturn and Pluto. Instructions for using a calculator are provided for each step in the Saturn example. The same operational sequences are used with the appropriate numbers for other planets.

Note: Although 1 AU = 92,955,807 miles, this has been rounded to 93,000,000 miles to simplify our examples.

Example: For Saturn, find T, given that d = 9.539 AU.
     From line (A) in the table, we find that:

$$T = [1.243325 \times 10^{-6}(9.539 \times 93)^3]^{\frac{1}{2}} = 29.46 \text{ years}$$

Press   [9][·][5][3][9][×][9][3][=][yˣ][3][=]
            [×][1][·][2][4][3][3][2][5][EE⌐][6][+/−][=][√x̄]

Display   29.46

From line (B) of the table, we find the value for K that gives the solution in terms of days:

$$T = [0.165863 \times (9.539 \times 93)^3]^{\frac{1}{2}} = 10,761 \text{ d}$$

Press $\boxed{9}\,\boxed{\cdot}\,\boxed{5}\,\boxed{3}\,\boxed{9}\,\boxed{\times}\,\boxed{9}\,\boxed{3}\,\boxed{=}\,\boxed{y^x}\,\boxed{3}\,\boxed{=}$

              $\boxed{\times}\,\boxed{0}\,\boxed{\cdot}\,\boxed{1}\,\boxed{6}\,\boxed{5}\,\boxed{8}\,\boxed{6}\,\boxed{3}\,\boxed{=}\,\boxed{\sqrt{x}}$

Display   **10761**

From equation (C), we find that:

$$T = (9.539)^{1.5} = 29.46 \text{ yr}$$

Press $\boxed{9}\,\boxed{\cdot}\,\boxed{5}\,\boxed{3}\,\boxed{9}\,\boxed{y^x}\,\boxed{1}\,\boxed{\cdot}\,\boxed{5}\,\boxed{=}$

Display   **29.46**

From equation (D), given that $T = 29.46$ years, we find that:

$$d_{AU} = 29.46^{0.6667} = 9.539 \text{ AU}$$

Press $\boxed{2}\,\boxed{9}\,\boxed{\cdot}\,\boxed{4}\,\boxed{6}\,\boxed{y^x}\,\boxed{0}\,\boxed{\cdot}\,\boxed{6}\,\boxed{6}\,\boxed{6}\,\boxed{7}\,\boxed{=}$

Display   **9.539**

From equation (E), given that $T = 29.46$ years and $d_{AU} = 9.539$, we find that:

$$\frac{T^2}{d_{AU}^3} = \frac{29.46^2}{9.539^3} = .9999009$$

Press $\boxed{9}\,\boxed{\cdot}\,\boxed{5}\,\boxed{3}\,\boxed{9}\,\boxed{y^x}\,\boxed{3}\,\boxed{=}\,\boxed{STO}\,\boxed{2}\,\boxed{9}\,\boxed{\cdot}\,\boxed{4}\,\boxed{6}\,\boxed{x^2}\,\boxed{\div}\,\boxed{RCL}\,\boxed{=}$

Display   **.9999009**

Example: For Pluto, find T, given that $d_{AU} = 39.526$.
  From (A), we find that:

$$T = [1.243325 \times 10^{-6}(39.526 \times 93)^3]^{\frac{1}{2}} = 248.5 \text{ yr}$$

From (B), we find that:

$$T = [0.165863 \times (39.526 \times 93)^3]^{\frac{1}{2}} = 90{,}766 \text{ d}$$

From (C), we find that:

$$T = (39.526)^{1.5} = 248.5 \text{ yr}$$

From (D), given that $T = 248.5$ years, we find that:

$$d_{AU} = 248.5^{0.6667} = 39.533 \text{ AU}$$

From (E), given that $T = 248.5$ years and AU = 39.533, we find that:

$$\frac{T}{d_{AU}^3} = \frac{(248.5)^2}{39.533^3} \approx 1$$

Application of Newton's Inverse-Square Law

Imagine that an object is dropped from a point 200,000 miles above the earth's surface. How long will it take to reach the earth's surface, and what velocity will it have achieved when it does? Both of these questions can be addressed using Sir Isaac Newton's inverse-square law.

First, let's look at uniform acceleration. Near the earth's surface, a falling object accelerates at approximately 32 feet per second (ft/sec^2). This means that it falls 16 feet the first second. Its velocity after falling from rest is 32 ft/sec at the end of the first second. Expressed mathematically, its average velocity for this first second is $(0 + 32)/2$, or 16 ft/sec. At the beginning of the second second, the velocity is 32 ft/sec and, being accelerated at 32 ft/sec^2, the object's velocity increases another 32 ft/sec during the second second to 64 ft/sec. Thus, the average velocity for this second second is $(32 + 64)/2 = 48$ ft/sec. The object has now descended a total of $16 + 48 = 64$ feet. This pattern repeats itself each second, meaning the object accelerates uniformly at the rate of 32 ft/sec^2.

Now, let's look at acceleration farther away from the earth's surface, where the rate is less than 32 ft/sec^2. For instance, let's investigate the acceleration of a falling object at a distance of 4,000 miles in space, or approximately 8,000 miles from the earth's center. The term *inverse-square* means varying inversely as the square of the distance from some origin. In this case, the acceleration is $(r_e/r)^2$ or $(4,000/8,000)^2 \times 32$, or 8 ft/sec^2.

Imagine how this rate of diminished acceleration affects our object dropped from 200,000 miles above the earth's center.

Using the same arithmetical approach, we find that the velocity at this point is $(4{,}000/200{,}000)^2 \times 32 = 0.0128$ ft/sec^2, significantly less than that at the earth's surface.

This is as far as we can go with the inverse-square law. To solve the rest of the problem, we need to apply the mathematics of integral calculus, which is beyond the scope of this book. We can, however, present the time-velocity relationship derived from the basic inverse-square law equations:

$$\frac{d^2s}{dt^2} = \frac{-k^2}{s^2}$$

This says that acceleration, or travel distance (s) per time unit (t) squared, is equal to the inverse square ($1/s^2$) of the distance multiplied by a constant (k).

By integrating and applying the appropriate constants of integration, we find that

$$v^2 = 2k^2 \left(\frac{1}{s_0} - \frac{1}{s_1} \right)$$

where s_0 is the earth's radius—3,963 miles or $(3.963 \times 10^3) \times (5.28 \times 10^3) = 20.925 \times 10^6$ feet—and s_1 is the point from which the object is dropped—200,000 miles or $(2 \times 10^5) \times (5.28 \times 10^3) = 10.56 \times 10^8$ feet.

The constant k^2 for the earth is equal to ar^2, or the earth's surface acceleration multiplied by the square of its radius. Thus, for earth's radius in feet, $k^2 = 14.097 \times 10^{15}$.

Substituting these values in our equation, we get:

$$v^2 = 2 \times 14.097 \times 10^{15} \left(\frac{1}{20.925 \times 10^6} - \frac{1}{10.56 \times 10^8} \right)$$

$$= 28.194 \times 10^{15} \left(\frac{0.047790}{10^6} - \frac{0.094697}{10^8} \right)$$

$$= (28.194 \times 10^{15})(0.046843 \times 10^{-6})$$

$$= 13.206924 \times 10^8 \text{ ft/sec}^2$$
$$v = 3.63413 \times 10^4 \text{ ft/sec}$$

Thus, the object's velocity is 36,341.3 feet per second, or (if we divide by 5,280 feet per mile) 6.88 miles per second.

To solve this equation in terms of miles per second using a scientific calculator, we first need to rearrange terms:

$$v = \frac{\left[28.194 \times \dfrac{10^{15}}{10^6} \times \left(\dfrac{1}{20.925} - \dfrac{1}{10.56 \times 10^2} \right) \right]^{\frac{1}{2}}}{5,280}$$

Now, we can proceed with the following sequence:

Press   [1] [÷] [1] [0] [·] [5] [6] [+/−] [EE↓] [2] [=] [STO] [1] [÷] [2] [0] [·]

           [9] [2] [5] [=] [+] [RCL] [=] [×] [2] [8] [·] [1] [9] [4] [EE↓] [9]

           [=] [√x̄] [÷] [5] [2] [8] [0] [=]

Display   **6.88**

It is interesting to note that about twenty-five thousand years ago a huge meteor hit the earth near Flagstaff, Arizona. The crater is approximately 0.75 miles in diameter and 650 feet deep. It is estimated that the meteor struck at a speed of approximately 6.84 miles per second, or 11 kilometers per second.

Going back to our original question, how long will it take the object to travel from 200,000 miles above the earth to the earth's surface? We can use the following formula

$$T = \frac{\pi}{k} \left(\frac{s_1}{2} \right)^{\frac{3}{2}}$$

where T is the travel time, k is the constant, and s_1 is the distance to be traveled (200,000 miles or 10.56×10^8 feet). Note that this formula actually reflects the time for the object to fall to $s_0 = 0$. The theoretically correct formula, taking into account the earth's radius ($s_0 = 3,963$ miles), is quite complex. If you are familiar with integral calculus, Appendix E gives the derivation of v and T.

Substituting the numbers we used earlier, we get:

$$T = \frac{\pi}{(14.097 \times 10^{15})^{\frac{1}{2}}} \left(\frac{10.56 \times 10^8}{2}\right)^{\frac{3}{2}}$$

$$= \frac{\pi}{11.873 \times 10^7}(12.1325 \times 10^{12}) = 3.2102563 \times 10^5 \text{ sec}$$

$$= 89.1738 \text{ hr} = 3.716 \text{ d}$$

To use a scientific calculator for this problem, use the following sequence for an answer in terms of days:

Press   `( ( ( 1 4 · 0 9 7 EE⁺ 1 5 √x STO )`
                 `( ( 1 0 · 5 6 EE⁺ 8 ÷ 2 = yˣ 1 · 5 = ) )`
                 `÷ RCL = × π ÷ 3 6 0 0 ÷ 2 4 =`

Display   3.716

Halley's Comet

Edmund Halley (1656–1742), the British mathematician and astronomer who financed the publication of Newton's *Principia Mathematica,* used Newton's theory to calculate the orbit of the great comet of 1682 (known ever afterward as Halley's Comet). He predicted that it would return in 1758, which it did. The comet reappears every 76 years or so, and records of its appearance have now been identified as far back as 2,200 years ago. The comet last appeared in February 1986 and is now speeding swiftly away from the sun. The most famous comet in history will again pass by the earth in the year 2062.

The orbit of Halley's Comet is an ellipse 36.18 AU (astronomical units) long by 9.12 AU wide. With the sun as a focus of the ellipse, at what point does the comet's path come closest to the sun? We begin with the following information

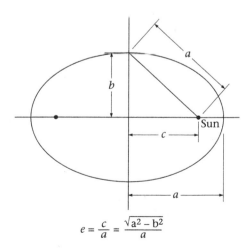

$$e = \frac{c}{a} = \frac{\sqrt{a^2 - b^2}}{a}$$

$$e = \frac{\sqrt{a^2 - b^2}}{a}$$

where $2a$ is the length and $2b$ is the width of the ellipse; e is the degree of departure from the circle. Plugging in our numbers we get:

$$e = \frac{\sqrt{18.09^2 - 4.56^2}}{18.09} = 0.96770823$$

Because $e = c/a$, c is equal to ea, so we can solve for c:

$$c = 0.96770823 \times 18.09 = 17.5058419 \text{ AU}$$

The closest point of the comet's path can then be calculated by $a - c = 18.09 - 17.5058419 = 0.5841581$ AU, or 54,326,700 miles.

Note: The observed perihelion in 1986 was 55,000,000 miles.

To solve this using a handheld calculator, the operational sequence is:

Press   [1] [8] [.] [0] [9] [x^2] [−] [4] [.] [5] [6] [x^2] [=] [$\sqrt{x}$]
[÷] [1] [8] [.] [0] [9] [=] [×] [1] [8] [.] [0] [9] [−]
[1] [8] [.] [0] [9] [=] [+/−] [×] [9] [.] [3] [EE↓] [7] [=]

Display   **54326700**

Einstein's Special Theory of Relativity and the Twin Paradox

In 1905, Albert Einstein published his special theory of relativity. No other scientific revelation since Newton's laws of gravitation over two hundred years earlier had such an impact! Throughout history, it had been taken for granted that time was absolute— that beings anywhere in the universe would measure time the same way humans did on earth. Einstein shattered this illusion.

The first step in his analysis was the discovery of the *relativity of simultaneity.* This meant that two events taking place at two distant locations may appear to one observer to occur at the same time, whereas another observer in a different state of motion sees them occurring at different times. Thus, the assumption of a common or universal time could no longer be justified on the premise that using a fixed procedure would allow anyone to establish such a time independent of his or her state of motion. Accordingly, two voyagers in space will not measure motion the same way, as distance *and* time are not absolutes. Rather, we must assume that each of us constructs our own frame of reference. Such frames of reference should consist not only of measured boundaries marking off specific locations in space, but also of clocks placed at the various locations. All of these elements then move synchronously together with the person involved. If a second person in a different state of motion viewed the clocks from the first person's frame of reference, he would find them unsynchronized (though, as perceived by each, all are ticking at the same rate). To the second person, the clocks located forward in the direction of the first frame's motion would appear relatively retarded, whereas those located backward in the direction of motion would seem advanced.

This apparent paradox is expressed mathematically by the *Lorentz transformation*:

$$ t' = \frac{\left(1 - \dfrac{v}{c}\right) \times t}{\sqrt{1 - \dfrac{v^2}{c^2}}} $$

The special theory of relativity formula relates to the time (t') for an object in space moving at a velocity of v miles per second, with respect to an observer's time at rest (t).

The speed of light (c) comes from the fundamental axiom of relativity, the premise on which Einstein's entire theory is based. The axiom says that the speed of light is absolute (always the same) regardless of the point of view from which it is measured. In free space, this speed is approximately 186,000 miles or 300,000 kilometers per second. Therefore, whether we measure the speed from earth, the moon, or a satellite, it will be identical.

Relativity theory also states that it is impossible to travel faster than the speed of light and that all forces and effects are limited to this speed. There is still no known way to transmit information via any faster medium.

The Twin Paradox

Let's look at a practical example to illustrate this theory. Jim and Luke are identical twins, but Jim is adventurous while Luke is cautious. When they are twenty, Jim takes an opportunity to travel to distant points in the galaxy in a rocket ship that can achieve velocities close to light speed for long periods of time. Luke decides to remain at home. From Luke's point of view, Jim ages at a diminished rate—sometimes at a greatly diminished rate. When Luke is sixty, Jim returns. However, because from

Luke's point of view Jim has been aging at a slower rate, Jim is only thirty when he returns.

What is the paradox? Because from Jim's point of view he is stationary while Luke is moving, and so Luke is aging at a slower rate. Consequently, if Jim is thirty years old when he returns, Luke must be even younger. This cannot be true.

The special theory of relativity states that bodies moving uniformly relative to each other will record (per their timepieces) moving at a constant speed and their timepieces will indicate the same time. In exploring the galaxy, Jim has been undergoing changes in velocity. He started from zero velocity on earth, accelerated to some velocity approaching the speed of light for space travel, slowed down (negative acceleration) as he reached another planet, accelerated again on leaving, and so on. Jim's clock is indicating a lag with respect to Luke's clock. In fact, if Jim's ship could achieve the speed of light, his clock would stop!

Luke is unaware of Jim's measurement of time; hence, we cannot call this a paradox. No paradox is involved because Jim has undergone several periods of acceleration during his life, whereas Luke has been inertial (moving uniformly) all the time. Both men have measured their time, but there is no reason to believe that the two should be the same. There is no universal time, because time is a route-dependent quantity.

Let's apply the Lorentz transformation to the twins. When Jim leaves the earth, both twins are 20 years old. Jim averages 164,000 miles per second for 10 years according to his clock. Therefore:

$$\frac{v}{c} = \frac{164,000}{186,000} \cong 0.88$$

$$t' = \frac{(1 - 0.88)t}{\sqrt{1 - 0.88^2}} \cong 0.25t$$

Now, Luke's clock records that Jim has been away 40 years, but Jim's clock records only 0.25×40, or 10 years. So, when the twins meet, Jim will sensibly be 20 + 10 = 30 years old, and Luke will be 20 + 40 = 60 years old.

The Lorentz transformation can also be computed with a calculator:

Press $\boxed{0}\ \boxed{\cdot}\ \boxed{8}\ \boxed{8}\ \boxed{x^2}\ \boxed{-}\ \boxed{1}\ \boxed{=}\ \boxed{+/-}\ \boxed{\sqrt{x}}\ \boxed{1/x}\ \boxed{\text{STO}}\ \boxed{1}\ \boxed{-}$
$\boxed{0}\ \boxed{\cdot}\ \boxed{8}\ \boxed{8}\ \boxed{=}\ \boxed{\times}\ \boxed{\text{RCL}}\ \boxed{=}$

Display 0.25

Go Figure!

Energy Forever?
Einstein's Equation

Albert Einstein (1879–1955) discovered that the mathematical forms of the laws of physics are invariant (constant) in all inertial systems (those at rest or in uniform motion). This leads to the assertion of the equivalence of mass and energy and of change in mass, dimension, and time with increased velocity.

Mathematically, the relationship between mass and the energy it carries is given by the formula $E = mc^2$, where E is the energy and m is the mass, in joules and grams, respectively. The constant in the equation is c, which is the speed of light, or 2.9979×10^8 meters per second. The term *mass* in this instance is not related to atomic weight, so a gram of air is equivalent to a gram of gold when equating energy to mass.

This energy-mass equivalence principle made possible both the atom bomb and nuclear power facilities. A hypothetical example will give you some interesting insight into the tremendous amount of energy carried by only a kilogram of mass.

Let's assume that 1 kilogram (kg) of "at rest" matter is converted completely to energy. What would be the resultant energy in joules? Given that 1 joule is equal to 2.7778×10^{-7} kilowatt hours (kWh), how many kWh would be produced, and how long would that keep ten 100-watt light bulbs burning (ten 100-watt bulbs equal a 1-kilowatt "load")?

Using our equation with $m = 1$ kg and $c = 2.9979 \times 10^8$, we find that:

$$c^2 = 8.9874 \times 10^{16} \text{ m}^2/\text{sec}^2$$

Then insert this value for c^2 into Einstein's equation:

$$E = 1 \times 8.9874 \times 10^{16} \text{ J}$$

Now convert to kilowatt-hours:

$$(8.9874 \times 10^{16}) \times (2.7778 \times 10^{-7}) = 2.4965 \times 10^{10} \text{ kWh}$$

For ten 100-watt light bulbs to consume 2.4965×10^{10} kWh, where 365 days/year $\times$ 24 hours/day = 8.76×10^3 hours/year, we use the following equation:

$$t = \frac{2.4965 \times 10^{10}}{8.76 \times 10^3} = 2.85 \times 10^6 \text{ yr}$$

Will we ever find a light bulb that lasts almost 3 million years?

Note: The almost universally accepted solution to $E = mc^2$ problems is to use the metric system (meters/second). However, for English system devotees, the solution to the light bulb problem is still simple, using $c = 9.835710564 \times 10^8$ feet/second and $m = 1$ pound:

$$E = (1)(9.835710564 \times 10^8)^2$$
$$= 96.741201 \times 10^{16} \text{ lb-ft}^2/\text{sec}^2$$

Convert to kilowatt-hours:

$$(9.67412 \times 10^{17}) \times (1.1705586 \times 10^{-8}) = 11.324124 \times 10^9 \text{ kWh}$$

Solve for t:

$$t = \frac{1.1324124 \times 10^{10}}{8.76 \times 10^3} = 1.2927 \times 10^6 \text{ yr}$$

Food for Thought

Before we leave our discussion of time and space, it seems appropriate to challenge your thinking with something even Einstein couldn't figure out.

Just before his death in 1955, Albert Einstein was searching for a *unified field theory,* a commonality between the forces that control elementary atomic particle behavior and the gravitational forces that control the motions of celestial bodies. The answer eluded him, as it has others since. It may be that the immense disparity between the forces that hold electrons to nuclei and the weaker interactions of gravitation makes comparison seem impossible.

To get an idea of the relative strengths of electrical and gravitational forces, consider the interactions between two electrons, the particles that carry the electrical current in storage batteries and electric wiring. To reduce their electrical interactions to the level at which the same electrons attract each other gravitationally at a distance of one hundredth of an inch, we would need to separate the two electrons a distance of fifty light-years. This is roughly ten times the distance between neighboring stars, or some thousand trillion ($1,000,000,000,000,000$ or 10^{15}) miles.

In terms of the interactions between elementary particles, gravitational forces are almost too weak to be imagined. Nevertheless, they alone determine the motions of planets, stars, and so on, because gravitation combines two characteristics that tend to reinforce its efforts involving large bodies. In contrast to the binding forces within atomic nuclei, whose reach does not extend even to distances the length of an atom's diameter, gravitational forces remain significant at large distances. Further, electrical forces within and between atoms can be attractive or repulsive, tending to neutralize each other for large, electrically uncharged bodies. All gravitational forces, however, are attractive; bodies invariably gravitate toward each other. Hence, gravitational forces and their effects are unique in their ability to affect the paths in which the components of the solar system revolve around the sun.

So, here's the challenging question: Is there, or will there ever be, a unified field theory that correlates both gravitational forces and those of elementary atomic particles?

Stephen Hawking, the world-renowned British physicist and perhaps the most brilliant mind since Einstein, said in his book *Black Holes and Universes* that there are at least three possibilities:

1. There is a complete unified theory.

2. There is no ultimate theory, but there is an infinite series of theories.

3. There is no theory.

What do you think?

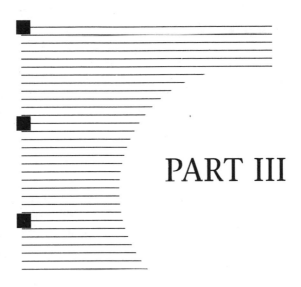

PART III

Did You Know?

For Neophytes

Did you know that . . .

....

- The lapsed time, in seconds, between seeing a flash of lightning and hearing a thunderbolt is a measure of the distance between the observer and the storm at the rate of 5 seconds per mile.

- The largest number expressed by three digits is $9^{9^9} =$ $2.95126 \ldots \times 10^{94}$.

- We could approach the speed of light (186,000 miles/second) if we could accelerate continuously through space at the rate of 1 g, where g is surface gravity, (32.2 feet/second^2) for 1 year.

- For any solid figure with polygonal faces, the number of faces (F) plus the number of vertices (V) less the number of edges (E) is always 2. This is expressed as $F + V - E$ and is called Euler's formula.

- Because of a tidal interaction causing the moon to recede from the earth, the earth's rotation is slowing 23.2 microseconds per year. This means that in 155 million years, there will be twenty-five hours in a day instead of twenty-four, and the year will have only 350.5 days.

- Every year that is divisible by 4 is a leap year, except for centennial years that are not divisible by 400. For example, the years 2000 and 2400 will be leap years, but 2100 and 2200 will not.

- Your heart beats over 2.2 billion times by the time you are 60 years old.

Go Figure!

- The length of a meter is approximately equal to the length of a pendulum that completes a single oscillation in one direction in one second.

- The universally accepted measure of a time unit—a second—is defined by the radiation frequency of the cesium 133 atom. This is the time required for the atom to vibrate 9,192,631,770 times.

- Betelgeuse, the supergiant star in the constellation Orion, has a diameter in excess of 80 AU (astronomical units), making it larger than 9,000 of our suns. It would fill our solar system past the asteroid belt.

- The nearest star system, Alpha Centauri, is 4.3 light-years (about 25 trillion miles) from the earth.

For Middle-of-the-Roaders

Did you know that . . .

• A number is divisible by 3 if the sum of its digits is divisible by 3.

• The smallest number expressed as the sum of the cubes of two numbers is 1,729 ($9^3 + 10^3 = 1,729$; $1^3 + 12^3 = 1,729$).

• Money invested at i percent interest will double in $0.30103 \div \log(1 + \frac{i}{100})$ years. For example, if you invest money at 8% interest, it will double in $0.30103 \div 0.03342 = 9$ years.

• A deck of fifty-two playing cards can be ordered 8.0658×10^{67}, or 52!, different ways.

• Every odd number greater than 7 is the sum of three prime numbers (Goldbach's second conjecture).

• Every number, other than a prime number, can be defined by a unique product of prime numbers.

• At a depth (h) in feet below the water surface, the pressure in pounds per square inch is $0.433h$.

• Were Bill Gates of Microsoft to invest his entire $56 billion today (June 1998) at 5% (after taxes), it would take him 40 years to exhaust his fortune spending $8,872,716 per day!

• The Pythagoreans of ancient Greece said an integer that was the sum of all its factors except itself was a *perfect number*. For example, $1 + 2 + 3 = 6$; $1 + 2 + 4 + 7 + 14 = 28$. The largest such number found to date is $2^{216,091} - 1$. Euclid

Go Figure!

(circa 300 B.C.) showed that if 2^{n-1} is a prime number, then $2^{n-1}(2^n - 1)$ is a perfect number.

- Fermat's "little theorem" states that $a^p - a$ is always divisible by p when p is a prime number. For example, $3^{11} - 3 = 177,144$, and $177,144 \div 11 = 16,104$.

- Every positive integer can be expressed as the sum of four or fewer squares of integers. For example, $1^2 + 2^2 + 3^2 + 4^2 = 30$; $1^2 + 3^2 + 5^2 = 35$. (The squares don't necessarily have to be different!)

- The expression $n^2 - n + 41$ generates prime numbers for all values of n from 1 to 40.

- A number is divisible by 8 if its last three digits are divisible by 8. For example, in the number 18,732,136, the last three digits (136) are divisible by 8 (17). Therefore, the entire number is divisible by 8 (2,341,517).

For Pure Math Buffs

Did you know that . . .

....

- A number is divisible by 2^n if the sum of the number's last n digits is divisible by 2^n. For example, the sum of the last three digits of $3{,}512 = 5 + 1 + 2 = 8$, for $n = 3$, and $2^3 = 8$, so $3{,}512$ is divisible by $2^n = 2^3 = 8$; $3{,}512 \div 8 = 439$.

- A number $x^n - 1$ is always divisible by $x - 1$ for all values of n. For example, $(7^4 - 1) \div 6 = (2{,}401 - 1) \div 6 = 400$.

- In exponential growth formulas, $e^{kx} \cong (1 + k)^x$ for $k \leq 0.015$ and $x \leq 85$ (maximum error <1%).

- An object falling below the earth's surface is attracted to the center by a force (F) equal to:

$$F = \frac{Gm_e}{r^2} = \frac{G\left(\frac{4}{3}\pi r^3\right)}{r^2} = \frac{r}{R_e} \times 31.197 \text{ ft/sec}^2$$

- If a number is divisible by a and b, and if a and b have no factors in common, the number is divisible by the product of a and b. For example, 861 is divisible by 3 and 7, and 3 and 7 have no common factors. Thus, $861 \div 21 = 41$.

- A multidigit number designated by $abcde$ is divisible by 11 if $a - b + c - d + e$ is divisible by 11. For example, if the number is 36,938, $3 - 6 + 9 - 3 + 8 = 11$ and $36{,}938 \div 11 = 3{,}358$.

Go Figure!

- Clifford A. Pickover, who edits "Brain Gogglers" in *Discover* magazine, reports on pairs of numbers called *vampires*. When two *x*-digit vampires are multiplied, they survive in a scrambled order as a 2*x*-digit vampire number. For example, $21 \times 87 = 1{,}827$. There are many larger vampire numbers. In fact, as of 1995, computers had established a world record with $1{,}234{,}554{,}321 \times 9{,}162{,}361{,}086 = 11{,}311{,}432{,}469{,}283{,}552{,}606$.

- A flag the area of *A* in square feet, furled in a wind with a velocity of *v* in miles per hour, exerts a force of *F* in pounds on the flagpole equal to:

$$F = 0.0003Av^{1.9}$$

- The weight (*W*) in pounds of a female African elephant at age *t* years may be approximated by the following formula:

$$W = 5{,}730(1 - 0.51e^{-0.075t})^3$$

where *e* is the base of natural logarithms and has a value of 2.71828. (See Appendix D for further discussion of logarithms.)

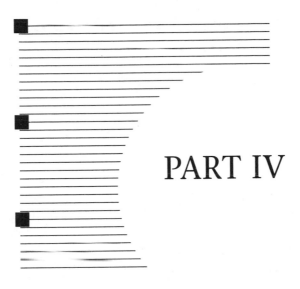

PART IV

One for the Road

Lazy Method of Substitution in Polynomials

There comes a time when every author knows he or she should call it quits. Then a topic comes to mind that is just too good to leave out. It doesn't fit in any of the established sections of the book, but it is still of interest to the reader. So, here's my last hurrah for tried-and-true math buffs.

Given $f(x) = 4x^4 - 14x^3 + 23x - 26$, what is the value for $x = 3$, or $f(3)$?

An easy way to find the answer is to create a table like the one shown on the facing page. (Interestingly, this procedure, known as *synthetic division*, is absent from a number of algebra textbooks.) First, arrange the coefficients of the terms in order of descending powers of x across the top row. Be sure to indicate whether each number is positive or negative, and insert a zero for any missing term (in this case, x^2). The constant term is also included. Now, beginning at the left, bring down the first coefficient, 4. In the second column, multiply 4 from the preceding column by 3; then add the product to the second coefficient. Multiply this result by 3; add the product to the third coefficient, and so on. The final result, –11, is the value of the polynomial when $x = 3$.

To check your answer, substitute 3 for x in the polynomial:

$$4(3)^4 - 14(3)^3 + 23(3) - 26 = 324 - 378 + 69 - 26 = -11$$

You can reverse the operation to find the value of the polynomial when multiplied by $x - 3$. Using our previous example, the quotient is $4x^3 - 2x^2 - 6x + 5$ and the remainder is -11:

$$4x^4 - 14x^3 + 23x - 26 = (x - 3)(4x^3 - 2x^2 - 6x + 5) - 11$$

4   **−14**   **0**   **23**   **−26**

$4 \times 3 = 12$

$12 - 14 = -2$

$-2 \times 3 = -6$

$-6 + 0 = -6$

$-6 \times 3 = -18$

$-18 + 23 = 5$

$5 \times 3 = 15$

$15 - 26 = -11$

4   −2   −6   5   −11

Appendix A

Metric Conversions

	Multiply	**By**	**To Convert To**
Mass	grams	2.2046226×10^{-3}	pounds
	kilograms	2.2046226	pounds
	pounds	4.5359237×10^{2}	grams
	pounds	0.45359237	kilograms
Length	centimeters	0.3937	inches
	meters	39.37	inches
	centimeters	3.28084×10^{-2}	feet
	meters	3.28084	feet
	kilometers	3,280.84	feet
	meters	6.2137×10^{-4}	miles
	kilometers	0.6213712	miles
	inches	2.54	centimeters
	inches	2.54×10^{-2}	meters
	feet	30.48	centimeters
	feet	0.3048	meters
	feet	3.048×10^{-4}	kilometers
	miles	1,609.344	meters
	miles	1.609344	kilometers
	kilometers	1.057023×10^{-13}	light-years
	miles	1.701114×10^{-13}	light-years

Multiply	By	To Convert To
light-years	9.460528×10^{12}	kilometers
light-years	5.878500×10^{12}	miles

	Multiply	By	To Convert To
	grams/centimeter^3	62.42796	pounds/feet^3
	grams/centimeter^3	3.612751×10^{-2}	pounds/inch^3
Density	pounds/feet^3	1.6018463×10^{-2}	grams/centimeter^3
	pounds/feet^3	1.601846×10^{-5}	kilograms/centimeter^3
	pounds/inch^3	27.679905	grams/centimeter^3

	Multiply	By	To Convert To
	kilometers/hour	54.68066	feet/minute
	kilometers/hour	16.66667	meters/minute
	kilometers/hour	0.277778	meters/second
Velocity	kilometers/hour	0.6213712	miles/hour
	miles/hour	1.466667	feet/second
	miles/hour	26.8224	meters/minute
	miles/hour	0.44704	meters/second

Easy Conversions

For feet per second (ft/sec) to miles per hour (mi/hr):

$$\frac{\text{ft}}{\text{sec}} \times \frac{1/5{,}280 \text{ mi/ft}}{1/3{,}600 \text{ hr/sec}} = \frac{3{,}600}{5{,}280} = 0.68182 \text{ mi/hr}$$

Example: Convert 88 ft/sec to miles per hour:

$$\frac{88 \text{ ft}}{\text{sec}} \times \frac{1/5{,}280 \text{ mi/ft}}{1/3{,}600 \text{ hr/sec}} = \frac{88 \times 3{,}600}{5{,}280}$$

$$= 88 \times 0.68182 \text{ mi/hr} = 60 \text{ mi/hr}$$

Go Figure!

For grams per cubic centimeter (g/cm³) to pounds per cubic feet (lb/ft³):

$$\frac{g}{cm^3} \times \frac{2.2046226 \times 10^{-3} \left(\frac{lb}{g}\right)}{(3.28084 \times 10^{-2})^3 \left(\frac{ft}{cm}\right)^3} = \frac{2.2046226 \times 10^{-3}}{3.5314667 \times 10^{-5}}$$

$$= 62.42796 \ lb/ft^3$$

Example: Convert 1 g/cm³ to pounds per cubic foot:

$$\frac{1 \ g}{cm^3} \times 62.42796 \cong 62.428 \ lb/ft^3 \text{ (density of 1 cubic foot of water)}$$

Appendix B

Scientific Notation

For a number of the problems in Part II, it is appropriate to introduce or review scientific notation, which is useful for calculating with very large or very small numbers. A few examples will demonstrate its usefulness.

The distance a ray of light travels in one year is approximately 5,900,000,000,000 miles. This number may be written using scientific notation as 5.9×10^{12}. The positive exponent 12 indicates that the decimal point should be moved 12 places to the right. The notation works equally well for small numbers. To illustrate, the weight of an oxygen molecule is estimated to be 0.00000000000000000000053 grams or, in scientific form, 5.3×10^{-23} grams. The negative exponent indicates that the decimal point is moved 23 places to the left.

Even more manageable numbers can be expressed this way:

$$513 = 5.13 \times 10^2$$
$$92,000,000 = 9.2 \times 10^7$$
$$0.00000000043 = 4.3 \times 10^{-10}$$

Many calculators employ scientific notation in their display panels; for the number $c \times 10^n$, the 10 is suppressed and only the exponent is shown. For example, to find $(4,500,000)^2$ on a scientific calculator, we would enter the number 4500000 and press the x^2 (or squaring) key. The display panel would show 2.025^{13} or 2.025 13. We would translate this as $4,500,000^2 = 20,250,000,000,000$.

Go Figure!

Appendix C

Formulas, Relationships, and Notation

Examples of Scientific Notation

- $20{,}250{,}000{,}000{,}000 = 2.025 \times 10^{13}$
- $0.00000000043 = 4.3 \times 10^{-10}$
- $92{,}000{,}000 = 9.2 \times 10^7$
- $0.000648 = 6.48 \times 10^{-4}$

Examples of n Factorial*

- $n! \approx \left(\dfrac{n}{e}\right)^n \sqrt{2\pi n}$
- $20! \approx 2.4329 \times 10^{18}$
- $50! \approx 3.0414 \times 10^{64}$

Examples of Exponential Functions

- $N(t) = N_0 e^{-0.20t}$ Decay of a cohort of Pacific halibut
- $N(t) = 248.3 e^{0.0076}$ Population growth
- $0(t) = 100{,}000 e^{-0.0558 \times .6}$ Decline in sales

---

*$\approx$ indicates "approximately equal to."

Typical Weather Formula (Windchill Factor)

$$T_{wc} = 91.4 - \frac{(10.45 + 6.686\sqrt{v} - 0.447v) \times (91.4 - T_f)}{22}$$

Typical Solar System Mechanics Formulas and Calculations

• Gravitational force: $F = \dfrac{Gm_1m_2}{r^2}$

• Centripetal force: $F = \dfrac{mv^2}{r}$

Kepler Relationships

• $T^2 = Kd^3; \quad K = \dfrac{4\pi^2}{Gm}$

• $T^2 = \dfrac{4\pi^2 \, d^3}{Gm}$

• $G = \dfrac{1.07}{10^9} \dfrac{ft^3}{lb. \ sec.}$

• $Gm_e = \dfrac{1.07}{10^9}(13.1 \times 10^{24}) = 14 \times 10^{15} \dfrac{ft^3}{lb. \ sec.}$

• $m_s = \dfrac{(899.94)(4.325 \times 10^5 \times 5.28 \times 10^3)^2 \times 10^9}{1.07}$

  $= 4.386 \times 10^{30} \, lb$

Exponentials and Radicals

- $a^0 = 1;$ $a^{-n} = 1/n$ ⟵ ⟲ ? $a^{-n} = \dfrac{1}{a^n}$; $2^{-3} = \dfrac{1}{2^3} = \dfrac{1}{8}$

Examples:

- $a^{m+n} = a^m \times a^n$ $4^{3+2} = 4^3 \times 4^2 = 64 \times 16 = 1{,}024$

- $a^{m-n} = a^m \times a^{-n} = a^m/a^n$ $4^3/4^2 = 64/16 = 4$

- $(a^m)^n = a^{mn}$ $(4^3)^2 = 4^6 = 4{,}096$

- $a^m/a^n = a^{m-n}$ $4^3/4^2 = 4^{3-2} = 4$

- $(ab)^m = a^m b^m$ $(4 \times 3)^3 = 4^3 \times 3^3$
$$= 64 \times 27 = 1{,}728$$

*• $a^m \times b^n$ No solution

- $a^n/b^n = (a/b)^n$ $4^2/3^2 = (4/3)^2 = 16/9$ or 1.778

- $a^{\frac{1}{n}} = \sqrt[n]{a}$ $27^{\frac{1}{3}} = \sqrt[3]{27} = 3$

- $a^{\frac{m}{n}} = \sqrt[n]{a^m}$ $27^{\frac{2}{3}} = \sqrt[3]{27^2} = \sqrt[3]{729} = 9$

*• $\sqrt[n]{ab} = \sqrt[n]{a} \times \sqrt[n]{b}$ $\sqrt[3]{27 \times 64} = \sqrt[3]{27} \times \sqrt[3]{64}$
$$= 3 \times 4 = 12$$

*• $\sqrt[n]{\dfrac{a}{b}} = \dfrac{\sqrt[n]{a}}{\sqrt[n]{b}}$ $\sqrt[3]{27/64} = \dfrac{\sqrt[3]{27}}{\sqrt[3]{64}} = \dfrac{27^{\frac{1}{3}}}{64^{\frac{1}{3}}} = \dfrac{3}{4}$

*• $\sqrt[m]{\sqrt[n]{a}} = \sqrt[mn]{a} = a^{\frac{1}{mn}}$ $\sqrt[2]{\sqrt[3]{64}} = \sqrt[6]{64} = 64^{\frac{1}{6}} = 2$

*• $\dfrac{1}{\sqrt[n]{b}} = \dfrac{1}{b^{\frac{1}{n}}} = b^{\frac{-1}{n}} = (b^{-1})^{\frac{1}{n}}$ $\dfrac{1}{\sqrt[3]{27}} = \left(\dfrac{1}{27}\right)^{\frac{1}{3}} = \dfrac{1}{3}$

---

*These are challenges and require a good understanding of exponentials and radicals.

Appendix D

Commentary: Logarithms

Although any positive number *b* larger than unity might have been chosen as a base of a system of logarithms, two numbers have actually been chosen in the construction of logarithmic tables: the number *b* = 10 and the number *e* defined as the limiting value of:

$$\left(1 + \frac{1}{n}\right)^n = e \approx 2.71828 \ldots$$

as *n* approaches infinity. The system of logarithms to the base 10 is usually referred to as *common logarithms;* the system of logarithms to the base *e* is called *natural logarithms.*

Common logarithms have certain obvious advantages not shared by natural logarithms. All numbers between 1 and 10 have logarithms between 1 and 2, and so on.

Because the number *e* = 2.718 . . ., the base of natural logarithms is irrational; that is, it cannot be expressed as a ratio of two integers (therefore, when it is expressed as a decimal, it involves an infinite number of decimals with no repeating groups). Thus, it might seem odd that it has been chosen as the base of a system of logarithms. The primary motivation for this choice lies in the fact that the solutions of numerous problems in applied mathematics are most naturally expressed in terms of an exponent of *e* (or e^x). Thus, the solutions of problems such as the equilibrium of a flexible cable, the transient flow of electric current in a circuit, and the disintegration of radioactive elements are expressed in terms of e^x.

The tabulation of the function e^x was an indispensable aid in obtaining the solutions of many physical problems, as illustrated by the following examples.

....

- If $x = a^y$, then $y = \log_a x$. If $y = \log_a x$ and $x = a^y$, then $x = a^{\log x}$

....

Examples for a = 10, *commonly called "base 10"*

....

- $\log_{10} 1 = 0$ because $10^0 = 1$

  Examples:

- $\log (a \times b) = \log a + \log b$

  $\log (8 \times 12) = \log 8 + \log 12$
  $= 0.903 + 1.079 = 1.982$
  or $10^{1.982} = 96 = 8 \times 12$

- $\log (a/b) = \log a - \log b$

  $\log (8/12) = \log 8 - \log 12$

  $= 0.903 - 1.079 = -0.176$

  or $10^{-0.176} = \dfrac{1}{10^{0.176}}$

  $= \dfrac{1}{1.5} = 0.667 = \dfrac{8}{12}$

- $\log a^b = b \log a$

  $\log (12^3) = 3 \log 12 = 3(1.079)$
  $= 3.237;$
  or $10^{3.237} \cong 1{,}728 = 12^3$

- $\log n = -a$; or $n = \dfrac{1}{10^a}$

  $\log n = -3$, or $n = 10^{-3}$

  $= \dfrac{1}{10^{-3}} = \dfrac{1}{1{,}000}$

- $\log (1/a) = -\log a$

  $\log \tfrac{1}{4} = -0.60206$
  $= -\log 4 = -0.60206$

- $1/a \log n = \log n^{1/a}$

  $\tfrac{1}{4} \log 12 = 0.26979$
  $= \log 12^{\frac{1}{4}} = 0.26979$

- $\log 10^n = n$; $\log 10^{-n} = -n$

- $\log 10^m/10^{-n}$

  $\log 10^m - \log 10^{-n}$
  $= m - (-n) = m + n$

....

Relationships between $\log_{10}$ a *and* $\ln_e$ a *(for base* e = 2.71828 . . ., $\log_e$ *is expressed* ln)*

....

- $\log_{10} e = \log_{10} 2.71828 = 0.434295$

  $\ln 10 = 2.302585$

  Therefore,

  $\ln 10 = 2.302585 = 5.302 \log_{10} e$

  and $\log_{10} e = 0.434295 = 0.188612 \ln 10$

- $\ln x = -a \quad \ln x = -1.20337 \quad x = e^{-1.20337} = \dfrac{1}{e^{1.20337}}$

  $$= \dfrac{1}{3.331} = 0.300$$

- $\log_{10} x = \ln x \times \log_{10} e$ $\log_{10} 12 = \ln 12 \times \log_{10} e$

                                                 $1.07918 = 2.48491 \times 0.43429$

                                                 $= 1.07918$

- $\ln x = \log_{10} x \times \ln 10$ $\ln 12 = \log_{10} 12 \times \ln 10$

                                                 $2.48491 = 1.07918 \times 2.302585$

                                                 $= 2.48491$

....

Go Figure!

Appendix E

Commentary: Newton's Inverse-Square Law

Both the scope and power of mathematics expanded exponentially through the seventeenth century. With the invention of the calculus, Isaac Newton was able to organize all data on earthly and heavenly motions into one system of mathematical mechanics, which encompassed a ball falling to earth and the movements of the planets and stars.

The development of the calculus is also attributed to the German mathematician Gottfried Wilhelm Leibniz (1646–1716). At first, Leibniz worked independently of Newton. Whereas Newton had concentrated on finding the derivatives of functions (e.g., instantaneous rates of change, maxima and minima, the inverse square process), Leibniz is primarily responsible for the recognition that limits of sums—integral calculus—can be obtained by reversing differentiation.

This appendix is included to demonstrate the process and application of the calculus as it pertains to the application of Newton's inverse-square law.

Attractive Force Varying Inversely as the Square of the Distance

For positions in the positive direction from the origin, the velocity decreases algebraically as the time increases whether the motion is toward or from the origin; therefore, in this region the acceleration is negative. Similarly, on the negative side of the origin, the acceleration is positive. Since $\frac{k^2}{s^2}$ is always positive, the right member has different signs in the two cases. For simplicity, suppose the mass of the attracted particle is unity. Then the differential equation of motion for all positions of the particle in the positive direction from the origin is:

(1)
$$\frac{d^2s}{dt^2} = -\frac{k^2}{s^2}$$

On multiplying both members of this equation by $2\frac{ds}{dt}$ and integrating, it is found that:

(2)
$$\left(\frac{ds}{dt}\right)^2 = \frac{2k^2}{s} + c_1$$

Suppose $v = v_0$ and $s = s_0$ when $t = 0$; then:

$$c_1 = v_0{}^2 - \frac{2k^2}{s_0}$$

On substituting this expression for c_1 in (2), it is found that:

$$\frac{ds}{dt} = \pm\sqrt{\frac{2k^2}{s} + v_0{}^2 - \frac{2k^2}{3s_0}}$$

If $v_0{}^2 - \frac{2k^2}{s_0} < 0$, there will be some finite distance s_1 at which $\frac{ds}{dt}$ will vanish; if the direction of motion of the particle is such that it reaches that point, it will turn there and move in the opposite direction. If $v_0{}^2 - \frac{2k^2}{s_0} = 0$, then $\frac{ds}{dt}$ will vanish at $s = \infty$; and, if the particle moves out from the origin toward infinity, its distance will become indefinitely great as the velocity approaches zero. If $v_0{}^2 - \frac{2k^2}{s_0} > 0$, then $\frac{ds}{dt}$ never vanishes; if the particle moves out from the origin toward infinity, its distance will become indefinitely great, and its velocity will not approach zero.

Go Figure!

Suppose $v_0{}^2 - \frac{2k^2}{s_0} < 0$ and that $\frac{ds}{dt} = 0$ when $s = s_1$. Then equation (2) gives:

(3)
$$\frac{ds}{dt} = \pm\sqrt{\frac{2}{s_1}}k\sqrt{\frac{s_1 - s}{s}}$$

The positive or negative sign is to be taken according to the particle it is receding from, or approaching toward (the origin). This equation can be written in the form

$$\frac{s\,ds}{\sqrt{s_1 s - s^2}} = \pm\sqrt{\frac{2}{s_1}}\,k\,dt$$

and the integral is therefore:

$$-\sqrt{s_1 s - s^2} + \frac{s_1}{2}\sin^{-1}\left(\frac{2s - s_1}{s_1}\right) = \pm\sqrt{\frac{2}{s_1}}\,kt + c_2$$

Since $s = s_0$ when $t = 0$, it follows that

$$c_2 = -\sqrt{s_1 s_0 - s_0{}^2} + \frac{s_1}{2}\sin^{-1}\left(\frac{2s_0 - s_1}{s_1}\right)$$

whence:

(4)
$$\frac{s_1}{2}\left[\sin^{-1}\left(\frac{2s - s_1}{s_1}\right) - \sin^{-1}\left(\frac{2s_0 - s_1}{s_1}\right)\right]$$

$$+ \sqrt{s_1 s_0 - s_0{}^2} - \sqrt{s_1 s - s^2} = \pm\sqrt{\frac{2}{s_1}}\,kt$$

This equation determines the time at which the particle has any position at the right of the origin whose distance from it is less than s_1. For values of s greater than s_1, and for all negative values of s, the second term becomes imaginary. That means that the equation does not hold for these values of the variables; this was indeed certain because the differential equations (2) and (3) were valid only for:

$$0 < s \leqq s_1$$

Suppose the particle is approaching the origin; then the negative sign must be used in the right member of (4). The time at which the particle was at rest is obtained by putting $s = s_1$ in (4) and is:

$$T_1 = -\frac{1}{k} \sqrt{\frac{s_1}{2}} \sqrt{s_1 s_0 - s_0{}^2} - \frac{1}{k} \left(\frac{s_1}{2}\right)^{\frac{3}{2}} \left[+\frac{\pi}{2} - \sin^{-1}\left(\frac{2s_0 - s_1}{s_1}\right)\right]$$

The time required for the particle to fall from s_0 to the origin is obtained by putting $s = 0$ in (4) and is:

$$T_2 = -\frac{1}{k} \sqrt{\frac{s_1}{2}} \sqrt{s_1 s_0 - s_0{}^2} - \frac{1}{k} \left(\frac{s_1}{2}\right)^{\frac{3}{2}} \left[-\frac{\pi}{2} - \sin^{-1}\left(\frac{2s_0 - s_1}{s_1}\right)\right]$$

The time required for the particle to fall from rest at $s = s_1$ to the origin is:

$$T = T_2 - T_1 = \frac{\pi}{k} \left(\frac{s_1}{2}\right)^{\frac{3}{2}} (5)$$

Appendix F

Decimal Equivalents for Fractions of a Year

	Jan	Feb	Mar	April	May	June	July	Aug	Sept	Oct	Nov	Dec
1	0.0027	0.0877	0.1644	0.2493	0.3315	0.4164	0.4986	0.5836	0.6685	0.7507	0.8356	0.9178
2	0.0055	0.0904	0.1671	0.2521	0.3342	0.4192	0.5014	0.5863	0.6712	0.7534	0.8384	0.9205
3	0.0082	0.0932	0.1699	0.2548	0.3370	0.4219	0.5041	0.5890	0.6740	0.7562	0.8411	0.9233
4	0.0110	0.0959	0.1726	0.2575	0.3397	0.4247	0.5068	0.5918	0.6767	0.7589	0.8438	0.9260
5	0.0137	0.0986	0.1753	0.2603	0.3425	0.4274	0.5096	0.5945	0.6795	0.7616	0.8466	0.9288
6	0.0164	0.1014	0.1781	0.2630	0.3452	0.4301	0.5123	0.5973	0.6822	0.7644	0.8493	0.9315
7	0.0192	0.1041	0.1808	0.2658	0.3479	0.4329	0.5151	0.6000	0.6849	0.7671	0.8521	0.9342
8	0.0219	0.1068	0.1836	0.2685	0.3507	0.4356	0.5178	0.6027	0.6877	0.7699	0.8548	0.9370
9	0.0247	0.1096	0.1863	0.2712	0.3534	0.4384	0.5205	0.6055	0.6904	0.7726	0.8575	0.9397
10	0.0274	0.1123	0.1890	0.2740	0.3562	0.4411	0.5233	0.6082	0.6932	0.7753	0.8603	0.9425
11	0.0301	0.1151	0.1918	0.2767	0.3589	0.4438	0.5260	0.6110	0.6959	0.7781	0.8630	0.9452
12	0.0329	0.1178	0.1945	0.2795	0.3616	0.4466	0.5288	0.6137	0.6986	0.7808	0.8658	0.9479
13	0.0356	0.1205	0.1973	0.2822	0.3644	0.4493	0.5315	0.6164	0.7014	0.7836	0.8685	0.9507

Go Figure!

14	0.0384	0.1233	0.2000	0.2849	0.3671	0.4521	0.5342	0.6192	0.7041	0.7863	0.8712	0.9534
15	0.0411	0.1260	0.2027	0.2877	0.3699	0.4548	0.5370	0.6219	0.7068	0.7890	0.8740	0.9562
16	0.0438	0.1288	0.2055	0.2904	0.3726	0.4575	0.5397	0.6247	0.7096	0.7918	0.8767	0.9589
17	0.0466	0.1315	0.2082	0.2932	0.3753	0.4603	0.5425	0.6274	0.7123	0.7945	0.8795	0.9616
18	0.0493	0.1342	0.2110	0.2959	0.3781	0.4630	0.5452	0.6301	0.7151	0.7973	0.8822	0.9644
19	0.0521	0.1370	0.2137	0.2986	0.3808	0.4658	0.5479	0.6329	0.7178	0.8000	0.8849	0.9671
20	0.0548	0.1397	0.2164	0.3014	0.3836	0.4685	0.5507	0.6356	0.7205	0.8027	0.8877	0.9699
21	0.0575	0.1425	0.2192	0.3041	0.3863	0.4712	0.5534	0.6384	0.7233	0.8055	0.8904	0.9726
22	0.0603	0.1452	0.2219	0.3068	0.3890	0.4740	0.5562	0.6411	0.7260	0.8082	0.8932	0.9753
23	0.0630	0.1479	0.2247	0.3096	0.3918	0.4767	0.5589	0.6438	0.7288	0.8110	0.8959	0.9781
24	0.0658	0.1507	0.2274	0.3123	0.3945	0.4795	0.5616	0.6466	0.7315	0.8137	0.8986	0.9808
25	0.0685	0.1534	0.2301	0.3151	0.3973	0.4822	0.5644	0.6493	0.7342	0.8164	0.9014	0.9836
26	0.0712	0.1562	0.2329	0.3178	0.4000	0.4849	0.5671	0.6521	0.7370	0.8192	0.9041	0.9863

continued

	Jan	Feb	Mar	April	May	June	July	Aug	Sept	Oct	Nov	Dec
27	0.0740	0.1589	0.2356	0.3205	0.4027	0.4877	0.5699	0.6548	0.7397	0.8219	0.9068	0.9890
28	0.0767	0.1616	0.2384	0.3233	0.4055	0.4904	0.5726	0.6575	0.7425	0.8247	0.9096	0.9918
29	0.0795	—	0.2411	0.3260	0.4082	0.4932	0.5753	0.6603	0.7452	0.8274	0.9123	0.9945
30	0.0822	—	0.2438	0.3288	0.4110	0.4959	0.5781	0.6630	0.7479	0.8301	0.9151	0.9973
31	0.0849	—	0.2466	—	0.4137	—	0.5808	0.6658	—	0.8329	—	1.0000

Go Figure!